MICROWAVE DEVICES

MICROWAVE DEVICES

Dr. James T. Coleman

Reston Publishing Company, Inc.
A Prentice-Hall Company
Reston, Virginia 22090

Library of Congress Cataloging in Publication Data

Coleman, James T.
   Microwave devices.

   Includes index.
   1. Microwave devices.   I. Title.
TK7876.C64   621.381'33   82-299
ISBN 0-8359-4386-0   AACR2

© 1982 by **Reston Publishing Company, Inc.**
A Prentice-Hall Company
Reston, Virginia 22090

10 9 8 7 6 5 4 3 2 1

Interior design and production: **Jack Zibulsky**

Printed in the United States of America

Contents

Preface, ix
Introduction, xii

Chapter 1 Early Microwave Generators and Detectors, *1*
  1–1 Introduction, *1*
  1–2 Early RF and Microwave Power Generators, *1*
  1–3 The Barkhausen-Kurz Oscillator (BKO) Mode, *2*
  1–4 Detailed Description of BKO Mechanisms, *3*
  1–5 The Evolution of M- and O-Type Devices, *5*
  1–6 The Evolution of Solid State Microwave Devices, *6*
    1–6.1 Some Familiar Development Problems, *6*
    1–6.2 The Transferred Electron Device (TED), *7*
    1–6.3 The Avalanche Mode Development, *7*
  1–7 General Comments, *8*

Chapter 2 Solid–State Diodes and Parametric Amplifiers
        (PARAMPS), *9*
  2–1 Introduction, *9*
  2–2 Diode Principles, *9*
    2–2.1 The Nature of Semiconductors, *9*
    2–2.2 The Semiconductor Junction, *11*
    2–2.3 The Junction as a Rectifier, *12*
    2–2.4 The Junction as a Variable Capacitor, *13*
  2–3 Parametric Amplifier (PARAMP) Principles, *16*
    2–3.1 A Simple Parametric Amplifier, *16*
    2–3.2 The Double–Tuned PARAMP, *17*
    2–3.3 The Manley–Rowe Equations, *18*
    2–3.4 Analysis of Three Types of PARAMPS, *20*
    2–3.5 The Double–Resonant PARAMP, *23*
  2–4 Examples of Parametric Amplifiers, *25*

Chapter 3 Bipolar and Unipolar Microwave Transistors, *26*
  3–1 Introduction, *26*
  3–2 Basic Bipolar Transistor Principles, *27*
  3–3 The Microwave Bipolar Device, *28*

3–4   The Microwave Unipolar (FET) Transistor, *31*
3–4.1 The Junction Field Effect Transistor (JFET), *31*
3–4.2 The Schottky–Barrier Field Effect Transistor (MESFET), *32*
3–4.3 Principles of Operation, *33*
3–5   Bipolar and Unipolar Transistor Performance Comparison, *37*

Chapter 4   Transferred Electron Devices (TEDs), *39*
4–1   Introduction, *39*
4–2   The Property of Negative Differential Resistance (NDR), *39*
4–3   The Nature of the Gunn Mode, *42*
4–4   Delayed-domain (LSA) Mode, *44*
4–5   Examples of Gunn Oscillators, *47*

Chapter 5   Avalanche and Barrier Injection Devices, *49*
5–1   Introduction, *49*
5–2   IMPATT Device Principles, *50*
5–2.1 Avalanche Phenomena, *50*
5–2.2 IMPATT Oscillation, *52*
5–2.3 Double Drift IMPATTs, *54*
5–3   IMPATT Applications, *55*
5–4   TRAPATT Diodes, *56*
5–4.1 Discovery of the Mode, *56*
5–4.2 TRAPATT Oscillators, *57*
5–5   BARITT Devices, *60*
5–5.1 Device Principles, *60*
5–5.2 Examples of Applications, *63*

Chapter 6   Solid State Microwave Amplifiers, *64*
6–1   Introduction, *64*
6–2   The General Amplifier, *64*
6–2.1 Four Terminal Amplifiers, *67*
6–2.2 Two Terminal Amplifiers, *68*
6–3   Bipolar and Unipolar Microwave Amplifiers, *70*
6–4   TED (Gunn) Amplifiers, *71*
6–4.1 Stabilizing Consideration, *71*
6–4.2 Noise Level, *73*
6–4.3 Intermodulation Characteristics, *73*
6–5   Avalanche Injection Amplifiers, IMPATT and TRAPATT, *75*
6–5.1 The TRAPATT Amplifier, *75*
6–5.2 The IMPATT Amplifier, *75*
6–6   The BARITT Amplifier, *78*
6–7   Solid–State Amplifier Design Considerations, *78*

Chapter 7   Plasma Wave Electronics, *81*
7–1   Introduction, *81*
7–2   Plasma in Nature, *81*
7–3   Basic Plasma Concepts, *82*
7–4   Waves in the Plasma, *86*
7–5   Growing Waves in the Aurora Borealis, *89*

7–6   Experimental Evidence for the Generation and Control of Longitudinal Plasma Waves, *90*
7–7   Possible Plasma Wave Devices, *94*
7–8   Insights Into the Future, *96*

Chapter 8   M–Type Crossed–Field Devices, 97
8–1   Introduction, *97*
8–2   M–Type Device Principles, *98*
8–3   Types of Magnetrons, *101*
8–4   The Magnetron ''Baking–In'' Procedure, *105*
8–5   The Forward–Wave Crossed–Field Amplifier (FWCFA), *108*
8–6   The M–Carcinotron (M–BWO) Device, *109*
8–7   Applications of M–Type Devices, *111*

Chapter 9   O-Type Parallel–Field Devices, 114
9–1   Introduction, *114*
9–2   O-Type Device Principles, *114*
9–2.1 Velocity Modulation Process, *116*
9–2.2 The Continous Interaction Modulation Process, *118*
9–3   Travelling–Wave Tube Mechanisms, *119*
9–4   Multi–cavity Klystron Tube, *121*
9–5   The Reflex Klystron, *122*
9–6   The Twystron Tube, *125*
9–7   The Backward–Wave Device, *126*
9–8   Applications of O–Type Devices, *128*

Chapter 10   Masers and Lasers, 129
10–1   Introduction, *129*
10–2   Quantum–Electronic Concepts, *129*
10–3   Processes in the Maser, *133*
10–3.1 The Components of a Maser, *133*
10–3.2 Two, Three, and Four Level Maser Schemes, *134*
10–4   Examples of Laser Types, *135*
10–4.1 Crystalline Solid–State Lasers, *135*
10–4.2 Gas Lasers, *136*
10–4.3 Liquid, or Dye, Lasers, *137*
10–4.4 Chemical Lasers, *138*
10–4.5 The Semiconductor Laser, *140*
10–5   Laser Applications, *143*
10–5.1 Present Laser Usage, *143*
10–5.2 Examples of Usage. *143*
10–5.3 Future Laser Developments and Applications, *146*

Chapter 11   Josephson Junctions, 148
11–1   Introduction, *148*
11–2   The Principles of Superconductivity, *149*
11–3   Brian Josephson's Discovery, *151*
11–4   The Basic Josephson Effects, *153*
11–4.1 The DC Josephson Effect, *153*
11–4.2 The AC Josephson Effect, *154*
11–5   Device Limitations, *155*
11–5.1 Frequency Limits, *155*
11–5.2 Switching Times, *155*
11–5.3 Other Limitations, *155*

11–6   Electronic System Implementation of the Junction, *156*
11–6.1 The Precise Nature of the Junction, *156*
11–6.2 Use as a Magnetic Field Detector, *156*
11–6.3 Flux Transformers and the Gradiometer, *159*
11–6.4 Use as a Voltmeter/Thermometer, *160*
11–6.5 Use as a Millimeter Wave Detector, *161*
11–6.6 The Broadband Bolometer, *163*
11–7   System Applications, *164*

Chapter *12*   High Power at Millimeter Wavelengths, or the Gyrotron Tube, *165*
12–1   Introduction, *165*
12–2   The Gyrotron Principle, *165*
12–2.1 General Description of the Mechanism, *166*
12–2.2 Detailed Description of the Bunching Mechanism, *168*
12–2.3 Basic Types of Gyrotrons, *169*
12–2.4 Comparison with Conventional Linear–Beam Counterparts, *171*
12–2.5 Relativistic Effects in the Interaction Space, *175*
12–2.6 Operation at Harmonics of the Gyrofrequency, *176*
12–3   Gyrotron Operating Principles, *177*
12–4   Gyrotron Applications, *179*
12–4.1 Commercial Potential, *179*
12–4.2 Military Applications, *180*
12–4.3 Propagation Limitations, *180*
12–5   Commercial Availability, *180*
12–5.1 Present, *180*
12–5.2 Future, *181*

Chapter *13*   Intense Relativistic Beam Devices (IREBs), *182*
13–1   Introduction, *182*
13–2   The IREB Device Principles, *182*
13–2.1 Extended Conventional Devices, *182*
13–2.2 Nonconventional Devices, *184*
13–3   State of the Art and the Effect of Power Conditioning, *186*
13–4   IREB Applications, *187*
13–4.1 Research Use, *187*
13–4.2 Commercial Applications, 187

Chapter *14*   Device Concepts for the Future, *188*
14–1   Introduction, *188*
14–2   The Extended Vacuum Device, *189*
14–3   Solid State Devices, *190*
14–4   The Free Electron Device (UBITRON), *191*
14–5   Superconducting Devices, *193*
14–5.1 Josephson Detectors and Generators, *193*
14–5.2 Indium Antinomide (In Sb) Detectors, *194*
14–5.3 Cavity "Ringing" Devices, *194*
14–6   Plasma Wave Devices, *195*
14–7   The Hybrid Devices, *198*
14–7.1 The Electron Cyclotron Maser (ECRM), *198*
14–7.2 The Plasma Cyclotron Maser, *199*
14–7.3 Solid–State Plasma Devices, *200*

Index, *201*

Preface

Microwave Devices is written to present the basic principles of current and developmental microwave and millimeter wave devices, and to describe their present and future practical applications. The book is written as an aid to the practicing electronic engineer and for all others interested in gaining an understanding of the electronic principles involved. The material includes a digest of the available technology in solid state and vacuum microwave and millimeter wave device principles, drawn both from U.S. and foreign literature.

This book is an outgrowth of a course taught by the author at the George Washington University Department of Electrical Engineering and Computer Sciences. From the outset, it became apparent that no single book was available to cover the breadth of material envisioned, and thus a new source book was needed to enhance this course offering from the Electrophysics Option of the Department. The course was designed to emphasize the physical and electronic principles behind each basic device, developing an understanding that will aid the engineer or electronics specialist in his present and future device applications. As the course developed, it became apparent that there were basic similarities between the devices, similarities that permitted a uniform analysis technique to be developed to aid in the analysis of the devices. Over a three year period, the presentation of these concepts to students with a variety of electronic backgrounds permitted the author to develop this unified approach to device description, analysis, and applications.

In the present work, the device concept is reduced to its simplest terms, with the basic electronic mechanisms illustrated by simple analogies. With these analogies clearly in mind, the author moves toward the conversion of the device concept into a workable device form. In the author's opinion, the best work in electronic technology is done when the operating principles (and limitations) are first firmly grasped and then the device is properly tailored to take advantage of these principles. With this logic, future devices can be envisioned on the basis of an understanding of the mechanisms involved, and the steps required to transform the electronic concept into an efficient, workable device become more apparent. To illustrate this process, a section at the end of each chapter is devoted to the practical implementation and applications of that particular device, within its known limitations.

Microwave Devices provides a logical approach to the principles of the many present day devices, providing an outline and description of the types of devices possible based on these concepts. Each chapter also includes descriptions of present day applications of the given device family, and, where possible, gives typical circuits for their use. In addition, each chapter includes a section on future applications of the device family, noting whether the basic device has achieved maturity or whether significant device improvements in the future could result in expanded applications.

It should be recognized that the material in this book covers a broad range in the history of microwave devices. Some of the devices described are presently mature in their development, while others are nowhere near an "off the shelf" status. A book of this nature normally does not consider future devices, however, a sufficient amount of information was available from the author's experience to permit him to make suggestions on the likely future direction of device development and implementation.

JAMES T. COLEMAN
Annandale, Virginia
December 1981

Introduction

Progress in the generation, detection, and amplification of electrical signals has been both methodical and rapid. A review of the chapter headings of this book shows the initial, evolutionary development of microwave devices, from the early wireless work of Hertz (1887) and Marconi to pre-World War II developments. Spark gaps were used in the decades before World War II to generate microwave power. These generated a wide band of frequencies. The early work of Hertz resulted in frequency generation in the range of 60 to $500MH_z$. Marconi also used Hertzian apparatus to generate $500MH_z$ signals. Marconi changed to very low frequency (VLF) experiments when he found that these signals were more effective in long distance radio transmissions. Thus, the early progress in microwave development at the turn of the century was thwarted until the late 1920s and early 1930s. The popular vacuum tube was not suitable for microwave generation and interest in frequencies above $200MH_z$ was confined to the laboratory.

With the invention of crossed–field (Hull, 1921) and parallel–field vacuum devices (Heil, 1935; Varian, 1939), it became possible to develop microwave radars for World War II use. The magnetron and Klystron tubes, as representative of these two family types, made it possible to generate high microwave pulse powers and to detect them at large distances. With the invention of solid state diodes and related devices, small, efficient generators and amplifiers became feasible at microwave frequencies. The generic families of solid state devices expanded so that they dominated the low power amplifier and generator fields. The chapter headings in this book illustrate the expansion in types of microwave solid state devices. In many instances the device turned out to be a two terminal device, which in turn, put the burden on the microwave circuit designer to develop circuits that could separate the input and output signals that are naturally mixed by the two terminal design. As the need for higher frequencies developed (due to spectrum crowding, etc.), the microwave device spectrum was forced toward the millimeter wave spectrum. Solid state devices are presently being driven into this region of the spectrum, but lo and behold, we find new versions of the vacuum device appearing to meet the challenge. The gyrotron vacuum device has appeared to fill the gap between the millimeter wave generator and maser/laser devices. (See Figure 14–4 for an illustration of this gap). The time will come when usable devices will fill the entire spectrum between microwaves and infrared wave lengths.

This work covers more that the conventional devices operating in the microwave and millimeter wave bands. It attempts to introduce to the reader the trend toward higher frequency devices in addition to the present ''off-the-shelf'' electronics. For example, the Josephson junction, while not used in present commercial equipment, will set new records in detection sensitivity and signal frequency generation in future commercial applications. When used in computers, it will result in unheard of computational speeds. (Plans for such computers are outlined in Chapter 11). When high power millimeter wave pulses are needed, the gyrotron vacuum tube, which has already set new world records in the generation of millimeter wave pulse powers, will be used. With the need for very high millimeter wave pulse power, the electron beam in a tube must travel at speeds near to that of light. These tube types are called Intense Relativistic Beam Devices. The electrons, travelling at nearly the speed of light, must follow the laws of relativity. They thus have their mass increased due to this very high energy and speed. This very increase in the electron's mass is used in the gyrotron tube to couple energy from its hollow electron beam into the circular waveguide surrounding it. Devices of the future are treated in the last chapter of this book, with emphasis upon the orderly development of these devices.

The development of microwave and millimeter wave devices has been both evolutionary and revolutionary. The discovery of oscillating and amplifying solid state devices was revolutionary. Nobel prizes were awarded for such work. The discovery of the Josephson effects (AC and DC) were also revolutionary and resulted in another Nobel prize. But the revolutionary discovery must be followed by careful development of practical devices based upon the new concept, and further followed by the development of practical circuits that can accept these devices. When the device–circuit work is mature, the systems must be revised to accept these new elements. Each chapter in this book covers the device concept, its practical implementation, and its circuit matching problems. Each chapter also suggests some potential system application of such devices.

It is the author's belief that some electronic systems have been developed without consideration of the limitations of the types of devices used in them. Yet, the choice of the type of device used in the system is critical—critical because the factors of reliability, cost, efficiency, potential for expansion, etc., are strongly influenced by such choices. As a central source of information, this book offers the reader basic information on the major microwave and millimeter wave devices used today and anticipated for the future in the hope that a more careful choice can be made for optimizing future electronic systems.

MICROWAVE DEVICES

1 Early Microwave Generators and Detectors

1–1 INTRODUCTION

Over the past four decades there has been a dramatic advance in the use of the radio spectrum. The spectrum had advanced from the 10 meter (30 MHz) limit to the present millimeter-wave bands (30-300 GHz). Wavelengths shorter than 10 meters were designated as ultrahigh frequency (UHF) in the latter 30's. Even today, German-made FM broadcast receivers mark the FM band (88 to 108 MHz) with the 'U' of 'UKW' after the German words for ultrahigh frequency (*ultra kurz welle*).

The generation of ultrahigh frequencies is not a modern phenomenon. Heinrich Hertz conducted many of his early (1887) experiments using frequencies in the 60 to 500 MHz range. He used spark gap generators to generate the UHF and microwave signals. His nonlinear detectors were nonlinear devices called *coherers*. He used parabolic reflectors to focus the radio beams for transmission and detection. Even Marconi, usually considered the father of commercial wireless telegraphy, used Hertzian apparatus on frequencies to 500 MHz. He soon, however, switched to wave lengths longer than 200 meters when he found that they were more effective at long distances. Thus, the early radio experimenters were aware of short electromagnetic waves (UHF and above), but left them for what they felt was a more promising future for long wavelengths.

1–2 EARLY RF AND MICROWAVE POWER GENERATORS

By 1930, laboratory workers had succeeded in generating frequencies as high as 75 GHz, but not in a manner suitable for commercial interest. The signals were generated as high harmonics of spark gap rf power generators. The spark gap is capable of generating a very broad range of frequencies. Thus, a microwave cavity can be excited at its resonant frequency by means of a spark gap, even though the spark has only a small portion of its energy in the microwave bands. The microwave cavity responds only to those components within the resonant band of the cavity. The spark gap microwave generator was able to generate signal energy in the microwave band, but it generated this signal inefficiently. It was thus not of commercial interest at the time.

In the decade preceding World War II, rf power was generated by one of three available technologies: spark gaps, Alexanderson alternators, and vacuum tubes. Only the spark gap was capable of generating microwave power, and this only at the expense of huge amounts of energy at sub-harmonically-generated frequencies. The Alexanderson alternator was a mechanical electrical alternator that produced frequencies in the VLF region. The frequency was determined by the speed of rotation and the number of poles. The alternator could be keyed for radio telegraphy by switching the current in the field winding. In 1916 Navy engineers at radio station NAA (then located in Arlington, Va.) succeeded in amplitude-modulating the output of an Alexanderson alternator. Mechanical problems limited the alternator to the VLF frequency range. Vacuum tubes of the post-World War I era did not operate at UHF or microwave frequencies, at least not when operated in the conventional tube modes. They were limited by three phenomena: lead inductance and interelectrode capacitance, transit angle effects, and gain-bandwidth product. Reducing the size of the electrodes would reduce interelectrode capacitance, but would also limit the power dissipation and output power. Separating the electrodes would decrease capacitance, but would also increase the electron transit time. With special electrode geometries, tubes were operated up to 450 MHz in World War II. Transit time effects would occur when the time for an electron to travel (transit) from cathode to grid became comparable to one cycle of the applied grid voltage. This transit time effect is a fundamental limitation in the design of microwave vacuum tubes.

1–3 THE BARKHAUSEN-KURZ OSCILLATOR (BKO) MODE

In 1920 a solution to the transit time problem was proposed by Barkhausen and Kurz (Germany). The Barkhausen-Kurz oscillator (BKO) used a special vacuum tube configuration to generate signals up to 700

MHz in frequency. Figure 1.1 illustrates the mode of operation of the tube. Instead of the conventional negative grid and positive plate, the tube was operated with a positive grid and negative plate. Thus, the grid was made positive with respect to the cathode and the anode. Electrons leave the cathode K and are accelerated toward the positive grid, many of them travelling through the grid structure. After passing the grid structure, they are repelled by the negative plate potential back to the grid region. The electrons thus tend to oscillate in elliptical paths around the grid structure. As the figure shows, the output microwave power is taken from the grid terminal.

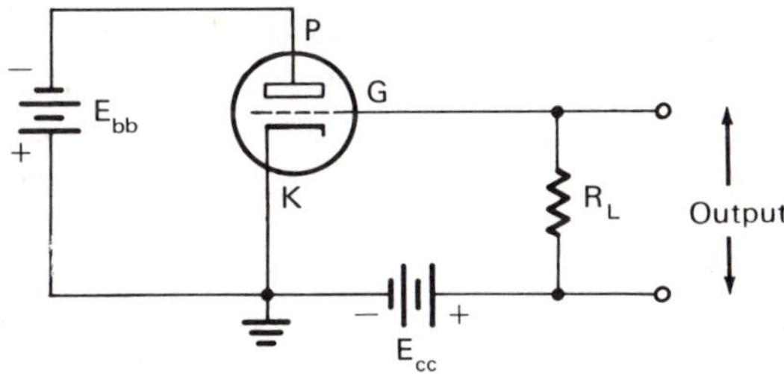

Figure 1-1: The Barkhausen-Kurz Mode of Triode Operation

American development of the BKO was hampered by an import embargo on non-American vacuum tubes during the radio patent wars of the 1920's. The BKO needed the cylindrical coaxial anode–grid configuration used in tubes in German design. The American tube design used flat, planar electrodes, a design not suitable for BKO operations.

The mechanism for the BKO mode of operation took advantage of the transit-time limitation, using it to build up electrical oscillation. Since the principles involved are basic to many of the other microwave tubes described in this book, a separate section is devoted to the BKO mechanism.

1–4 DETAILED DESCRIPTION OF BKO MECHANISMS

In order for the BKO mode to be sustained, the electrons must move in a nonrandom, synchronous manner. Consider an alternating voltage V_g superimposed on the grid bias voltage E_{cc}, with a frequency approximately equal to the reciprocal of the cathode-plate transit time. Consider an electron leaving the cathode as V_g is starting the negative swing of its rf cycle. The electron is still attracted to the grid, but with less acceleration due to the negative swing of V_g. As the electron passes the grid plane, the polarity of V_g reverses, and thus the electron is subject to an additional braking force and is stopped before it reaches the plate. The grid voltage V_g now reverses as the electron travels back toward the grid, and the electron has less acceleration than E_{cc} alone could provide. Upon the electron's reaching the grid, the polarity reverses again and the electron heads toward the cathode, where it is slowed down again by V_g. The net result is that the electron is constantly slowing down due to V_g, which means that the electron is giving up energy to the grid. The kinetic energy of the electron is being translated into oscillation energy. The electron is eventually captured by the grid.

Consider now an electron leaving the cathode in Fig. 1.1 at the time V_g is starting its positive rise. The electron gains additional energy as it moves toward the grid. As it passes the grid, it gains even more, due to V_g's reversal of polarity. The electron now may even reach the plate. As it traverses its path, it absorbs energy from V_g and thus tends to reduce the oscillation amplitude of V_g. The result is that these particular electrons do not remain in the interaction region near the grid. They will be soon drawn away. This sorting-out process permits the synchronous electrons to remain near the grid a much longer time than those electrons which receive energy from the grid. This means that there is a net resultant energy to sustain the electrical oscillations. Most electrons, then, that remain in the tube are synchronous and oscillating together (i.e., in phase).

Other factors need to be considered for a more exact analysis. The local space charge at the cathode needs to be considered, plus the tendency to collect some vibrating electrons. Another effect also occurs: the vibrating electrons that are out of phase with V_g tend to gain in phase, similar to the bunching action in the klystron tube. This latter action is described in detail in Chapter 9.

The BKO mode demonstrated that the electron transit-time effect could be put to use if considered properly, making it possible to use a conventional vacuum tube (preferably coaxial in design) to generate microwave oscillations. The efficiencies of microwave power generation are generally low, ranging from 1 to 3 percent for typical favorable conditions. Power output was also low, because the grid had to absorb the lost power and grids are usually not capable of dissipating large amounts of energy. Figure 1.2 illustrates one possible circuit for the generation of Barkhausen oscillation. In practice the BKO operated with a slightly negative plate and with a cathode with limited emission, so that the tube operated midway between a space-charge-limited and temperature-limited space current.

Other special modes of BKO operation have been developed, including tuning of the tank circuit at harmonics of the basic electronic

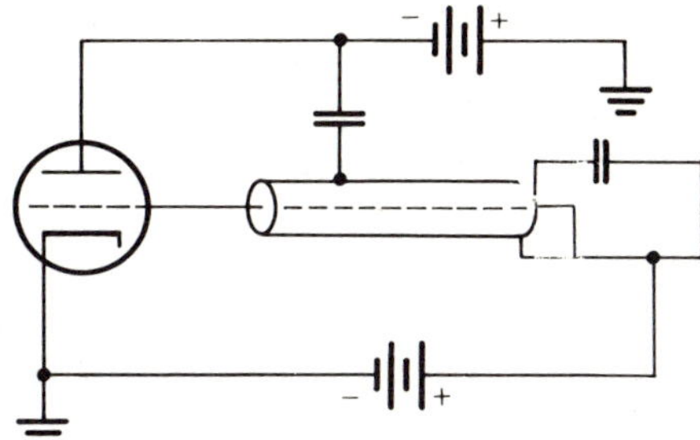

Figure 1-2: Coaxial Line Resonator Barkhausen Oscillator

oscillation. One mode used a resonance developed via the grid spiral structure.

The Barkhausen-Kurz oscillator served as a predecessor of the present microwave power generating tubes. In the work to follow, the basic electronic mechanisms of the BKO will be shown to serve as predecessors of many present day microwave devices.

1–5 THE EVOLUTION OF M- AND O-TYPE DEVICES

As mentioned previously, the BKO was severely limited in its power generation capability by the power dissipation capability of the triode's grid structure. (In some BKO experiments the grid would run white-hot during operation.) A solution to this problem was proposed in 1921 by A.W. Hull: delete the grid altogether, and keep the the electrons in orbit using a magnetic field. Hull's original magnetron has been much modified over the years, but the basic principle is still used today. Most microwave ovens, for example, use magnetron oscillators. One of the first modifications to the Hull magnetron was made by Yagi and Okabe working in Japan. The Yagi-Okabe magnetron achieved greater output power and higher operating frequencies by splitting the Hull anode into two or more sections. There were, however, still considerable design problems to solve because the small dimensions of the components required for microwave operation made the output power somewhat limited. By the mid-30's, however, Cleeton and Williams, working at the University of Michigan, achieved operation at 50 MHz. (Over the years, power and frequency have increased, making the magnetron one of the primary sources of microwave energy.) The magnetron was the first M-type microwave device developed.

The achievement of high power at higher frequencies remained a problem for several years. But in the mid-30's, several investigators simultaneously reached similar solutions. Dr. W.W. Hansen (Stanford University) and Drs. A. and O. Heil began to think in terms of turning the transit time to advantage through the mechanism of velocity modulation of the electron beam. The Drs. Heil proposed, in 1935, to use the electron transit time to control the electron stream. The heating problem was not solved, but was avoided, because the electrons would not actually strike the control electrodes.

Russell and Sigurd Varian extended Dr. Hansen's work into the practical world in 1937 when they used Hansen's calculations to build the first klystron vacuum tube. This device used the transit time, together with the deceleration of bunched electrons, to generate microwave rf energy. Velocity modulation of the electron stream in the klystron produces the bunching effect. It is the time between the arrival of successive bunches at a collector anode that determines the operating frequency of the klystron. Arrival of each bunch represents one cylce of rf energy. The klystron was the first O-type microwave device developed.

1–6 THE EVOLUTION OF
SOLID-STATE MICROWAVE DEVICES

1–6.1 Some Familiar Development Problems

The need for expansion into the microwave bands resulted in the development of many vacuum devices for power generation, amplification, and detection. The development of semiconductor devices saw similar, if not identical, problems. The high-frequency response of bipolar transistors, for example, was limited by the transit time of charge carriers (electrons or "holes") across the base region. Attempts at reducing the width of the base region in order to decrease transit time produced additional problems, i.e., increased capacitance and decreased tolerance to reverse bias potentials. But even thin base regions could not solve the problem. In semiconductor materials we see a property called electron saturation velocity, which is analogous to a similar property in vacuum tubes. This property seems to be a fundamental limit to the high frequency of bipolar transistors. But, just as in the case of the vacuum tube transit-time problem, we can turn the limitation into an advantage and use it to create microwave oscillations.

1–6.2 The Transferred Electron Device (TED)

John Gunn of IBM was studying the properties of n-type gallium arsenide (GaAs) material in 1963. He noticed that current passed through the material would become unstable if the applied voltage were increased above a certain threshold potential. It was found that the current would pulsate at microwave frequencies if the electric field exceeded the particular threshold value. Gunn had discovered a negative resistance effect that could be used to design microwave oscillators. It was later realized that electrons in the solid state material could exist in one of two conduction bands, and that higher-energy electrons would transfer into the higher conduction band that produced a higher effective electron mass. The drop in net electron velocity with increasing applied potential gave a negative resistance region in the volt-ampere characteristic of the GaAs material. The term *transferred electron device* (TED) refers to a device that transfers the electrons in a solid state material from one conduction band into another, usually higher, conduction band. Microwave oscillators that depend upon the transfer of electrons between high- and low-mobility conduction bands are called *transferred electron oscillators* (TEOs). Negative resistance oscillators were also the subject of speculation by Shockley as early as 1954. Esaki had produced a two-terminal diode-like device with a negative resistance property in 1957. Sommers suggested in 1959 that the Esaki diode could have microwave applications, a prediction that has proven to be accurate. Thus, the Gunn device was discovered against a background of previous negative resistance oscillator speculation and discovery.

The Gunn device operates in two modes, the transit-time mode and the limited space charge mode. In the transit-time mode, the thickness of the active region determines the oscillator frequency. In the limited space charge mode, the external tank circuit determines the frequency of operation.

1–6.3 The Avalanche Mode Development

In 1958 W.T. Read of the Bell Laboratories proposed the IMPATT (Impact Avalanche Transit Time) diode. Read's suggestion was that the phase delay in a p-n junction diode between an applied rf voltage and an avalanching current could be used for negative resistance operation at microwave frequencies. Fabrication problems prevented the construction of a Read diode until the mid-1960's. In 1965 R.L. Johnson of the Bell Laboratories verified Read's model experimentally when he generated 80 milliwatts of power at 12 GHz from a silicon p-n junction diode. It is now recognized that the Read structure is just one of many that will result in IMPATT operation.

In the IMPATT mode, the device is reverse biased near the avalanche point of the volt-ampere characteristic. The point is selected so that a small added potential can throw the device into the avalanche region. An oscillating tank circuit is placed across the diode biased in this state, and the tank oscillator will cause the diode to go into the avalanche mode during peaks of the rf cycle. The junction will exhibit negative resistance if the rf current can be made to lag behind the voltage by 90 degrees or more.

1–7 GENERAL COMMENTS

With this historical background, the work that follows will prove to form a familiar pattern of development. What appears to be a formidable limitation is often turned to an advantage. The limitation itself can often be exploited to form a new device based upon the so-called limitation, period. We will attempt to find commonalities in this work that could be useful in the exploration of new devices. The understanding of the basic mechanism of operation of each device is essential to this type of analysis. Often, the previous experience with one device can offer important insight into future device development. An obvious example is the parallel between vacuum tube and transistor development. Both had upper frequency limits based upon similar limitations. The transfer between similar technologies can aid in the development of a new device. Putting it another way, the history of device development should not be ignored, since in all of science we tend to build upon another individual's work.

2 Solid-State Diodes and Parametric Amplifiers (PARAMPS)

2–1 INTRODUCTION

Semiconductor devices have had a large impact upon the field of electronics. In fact, they have had a revolutionary impact. Before their advent, vacuum tubes dominated the electronic fields of rectification, amplification, detection, frequency multiplication, and signal generation. Vacuum tubes still dominate the fields of high-power microwave generation (for both peak and average powers). The devices of Chapters 8 and 9 are still dominant in this field. Figure 9-15 shows comparative performance for the vacuum reflex klystron and the solid-state Gunn effect diode oscillator, for frequencies above 3 GHz. Transistors, however, are not competitive above 3 GHz. Special semiconductor devices have been devised to extend the oscillation frequency to over 40 GHz, but these devices are generally two-terminal devices.

In this chapter, the basic semiconductor diode is discussed, and the exploitation of one of its properties for parametric amplification is analyzed. Once again, what at one time was considered to be a detrimental property proved to be an advantage. With this positive approach, we will find that the stumbling block can prove to be an advantage when it is properly understood and applied.

2–2 DIODE PRINCIPLES

2–2.1 The Nature of Semiconductors

The properties of a semiconductor depend strongly upon the energy band structure and the impurity levels injected into the semiconductor. First, it is necessary to consider the difference between a metal, an insulator, and a semiconductor material. Figure 2-1 compares the energy bands for those three cases.

Figure 2-1 shows how the conduction band in an insulator is separated from the valence band. The gap separation E_g is quite large, so that few electrons e$_-$ can ever move from the valence band into the conduction band to produce a flow of current in the insulator. In Fig. 2-1b the energy diagram for a semiconductor is given. The gap separation energy E_g is now small enough that at room temperature some electrons will have enough energy to cross the gap and current will flow when a potential is applied. The current will not be large, and therefore the material is called a semiconductor. In Fig. 2-1c, the energy diagram for a conductor (metal) is given. The conduction and valence bands overlap, and thus electrons can move with ease from the valence to the conduction bands, resulting in high conductivity in the material.

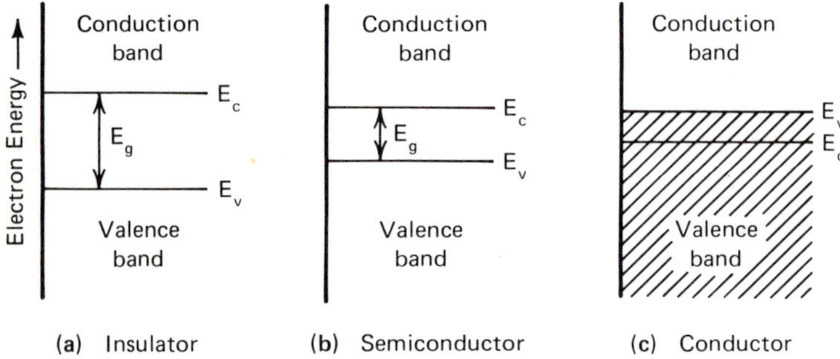

Figure 2-1: Energy Levels for Various Classes of Materials

An intrinsic semiconductor is one in which the action described above occurs at room temperatures. When certain foreign materials (dopants) are added to a semiconductor, the material can conduct electrons more easily or can even have conduction via a movement of electron vacancy sites ("hole current"). The doping additives can cause either electron current or hole current to flow. Those dopants that cause the electron flow create the n-type semiconductor, and those that create conditions for the flow of hole current result in p-type materials. Figure 2-2 illustrates the band structures necessary for the creation of n- and p-type semiconductor materials.

In Fig. 2-2a, electrons can move from E_e (dopant level) to the conduction band E_c with only the energy difference, E_c-E_e. Similarly, the energy required for hole conduction in Fig. 2-2b is the difference between E_n and E_v. The directions of increasing electron and hole energies are indicated in the figure.

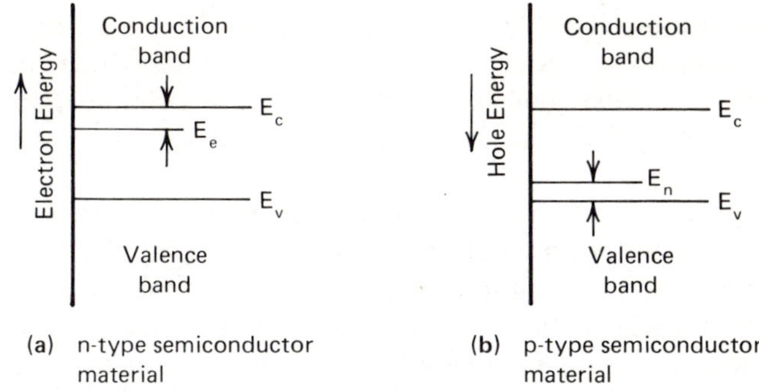

(a) n-type semiconductor
material

(b) p-type semiconductor
material

Figure 2-2: P- and N-type Semiconductor Energy Bands

2–2.2 The Semiconductor Junction

When an n-type semiconductor material is joined to a p-type material, a semiconductor junction is formed. The result is a joining of the energy bands of the two materials. The junction produces a kind of balance between the two energy diagrams of Fig. 2-2, the balance point (fulcrum) being the Fermi levels of the two materials. The energy diagram of Fig. 2-3 illustrates this. Note how the alignment of the two Fermi levels displaces the conduction and valence bands.

In the figure, there is a distance over which the energy bands are sloping. Outside this region, the effect of the junction discontinuity is not seen by charge carriers. Inside this region, electric fields exist that sweep away all charge carriers. That is, the electrons and holes are swept out of the region, leaving the uncompensated-for ions in the region of the abrupt junction.

The point l_o is the physical junction of the n– and p–type materials. The distance, $(l_n - l_p)$, is called the width of the junction, over which the

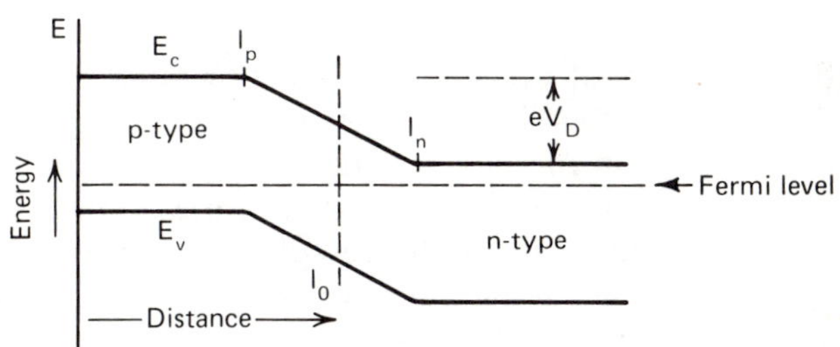

Figure 2-3: Abrupt P-N Junction

energy bands slope. The distances $(l_n - l_o)$ and $(l_o - l_p)$ are distances into the n– and p–type materials, respectively, where electric fields exist. There is a voltage present at this junction due to this charge displacement. The voltage is called the *diffusion potential* (V_D). Figure 2-4 shows the conditions at the junction.

The figure presents an interesting case, because there is now a potential across the junction without any voltage being applied. The uncompensated-for (donor) ions cannot be swept away, as they are fixed in their positions. As Fig. 2-4a shows, there is a positively charged region at the n side of the junction and a negatively charged one at the p side of the junction. Because of the existing electric field, shown in Fig. 2-4b, electrons are swept out of the inside of the junction, leaving the aforementioned uncompensated-for donor ions.[1] Also, the holes are swept out of the p side of the junction due to the field, leaving only the uncompensated-for acceptor ions. This results in a double layer of charges, positive and negative, as indicated in Fig. 2-4a.

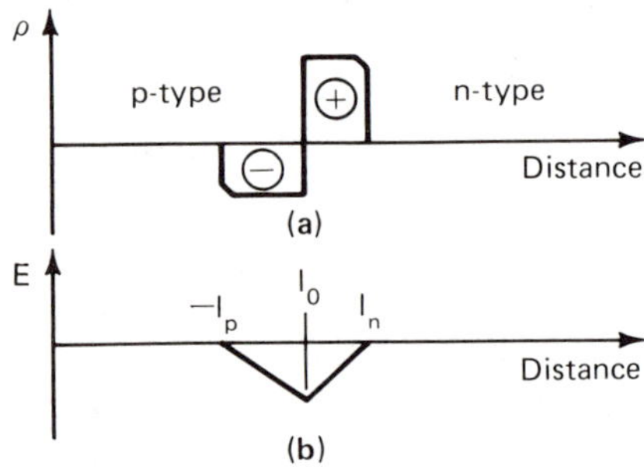

Figure 2-4: Charge and Field Distributions for an Abrupt P-N Junction

2–2.3 The Junction as a Rectifier

The diffusion potential V_D across the junction gives the electric field plot of Fig. 2-4b. If a voltage is placed across the junction that is of the same polarity as V_D, then the new junction energy diagram will appear as in Fig. 2-5. Note now that there is a bigger barrier voltage between n- and p-type materials, with the result that few charges can cross the barrier, and little current can flow. If we reverse the process, then V_D can be cancelled out, and the diode will conduct current. In order to do this, it

---

[1]The positive, donor ions form the n-type material, and the negative, acceptor ions form the p-type material.

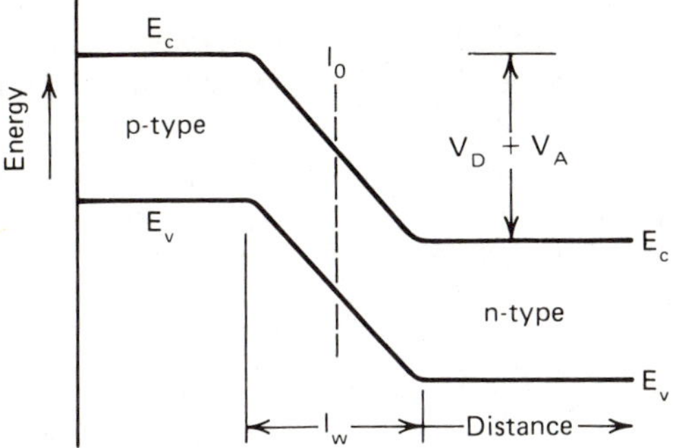

Figure 2-5: P-N Junction with Reverse Potential

is necessary to apply a potential equal to or greater than, but opposite in polarity to, the diffusion potential V_D. But this is no more than to apply what is called the *rectification property* of a p-n junction diode. The forward current direction (or high current direction) results when a potential exceeding, and opposite in polarity to, V_D is applied to the junction. This effect is illustrated in Fig. 2-6. The output voltage is equal to the peak generator voltage E_g minus the potential V_D during the positive generator cycle, and is nearly zero during the negative generator voltage cycle. It cannot be zero during the negative cycle of E_g, as a small reverse diode current will flow. There is a front-to-back diode resistance ratio that is quite high for most junction diodes, but since the back resistance is not infinite, then some small current will flow and a small potential will exist across R_1 during the negative E_g cycle (i.e., there aren't any perfect diode rectifiers).

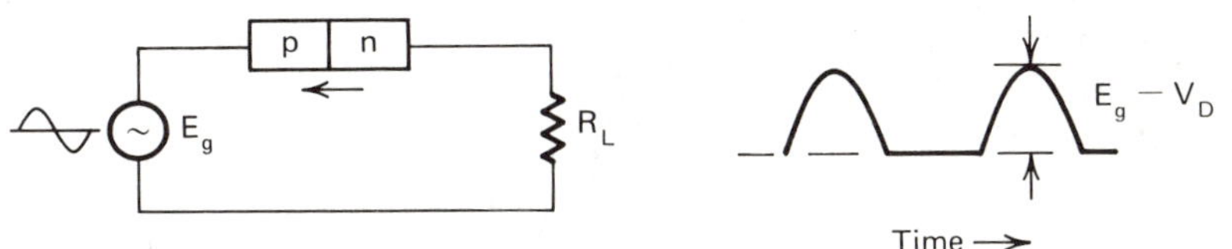

Figure 2-6: P-N Diode Rectification

2–2.4 The Junction as a Variable Capacitor

Referring back to Fig. 2-5, it should be noted that the width l_w between the p and n materials is sketched as wider than the equivalent width in Fig. 2-3. The relationship will be as in Fig. 2-7. The layer width W will be the value of $(l_n - l_p)$ as in Fig. 2-3, with no external potential applied. This is shown in Fig. 2-7a. With an external potential that opposes V_D, the case shown in Fig. 2-7b, the layer width will be less than W. With a value that aids V_D, as in Fig. 2-7c, the layer with l_w will exceed W. With this in mind, we see that the width of the charge layer will be modulated by the application of an external voltage.

In the case of a simple parallel plate capacitor, its capacitance is given by

$$C = \epsilon \frac{A}{d},$$

where ϵ is the dielectric constant, A is the plate area, and d is the separation between the plates. Decreasing the separation d will increase capacitance, and increasing the separation d will decrease the capacitance. It is clear, then, that the p-n diode will have its capacitance modulated as the applied potential reverses polarity. This effect can be used to produce a parametric amplifier, using the diode's variable capacitance as

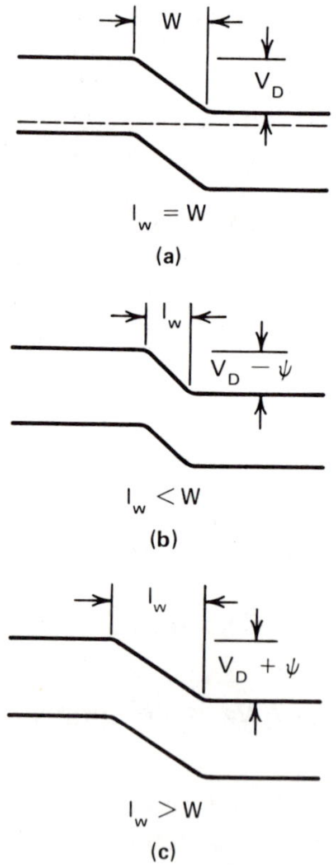

Figure 2-7: The Effect of External Potential on Layer Width

a circuit element. The diode can be designed to enhance this effect, and thus, diodes have been developed that perform best in a parametric amplifier or some other PARAMP applications.

The abrupt p-n junction is difficult to achieve in practice. Most junctions have some type of more gradual transition from the n- to the p-type material. Linear graded junctions have been produced by solid-state diffusion of the impurities into the bulk semiconductor materials. The results can be a gradual change from n- to p-type in a linear fashion. Figure 2-8 shows the impurity distribution in such a junction.

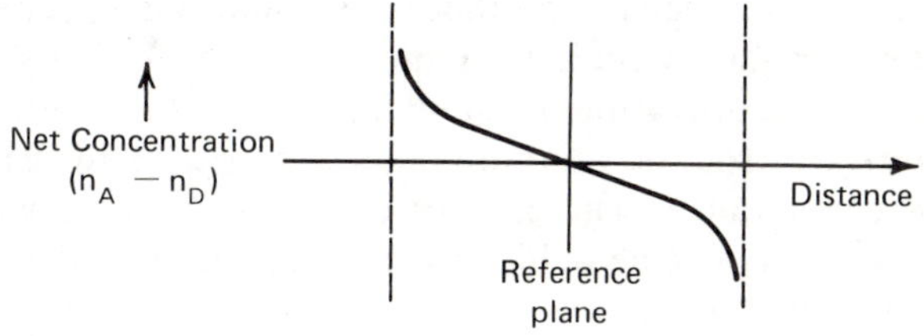

Figure 2-8: Impurity Levels in a Linear Junction

For the above two cases of junction conditions, the variation of junction capacitance with applied voltage has been computed. Figure 2-9 shows, in the form of a graph, how the two cases compare. The abrupt junction shows the greatest sensitivity to a change in the junction voltage V. (The formulas for the two cases appear above the curves in Fig. 2-9.)

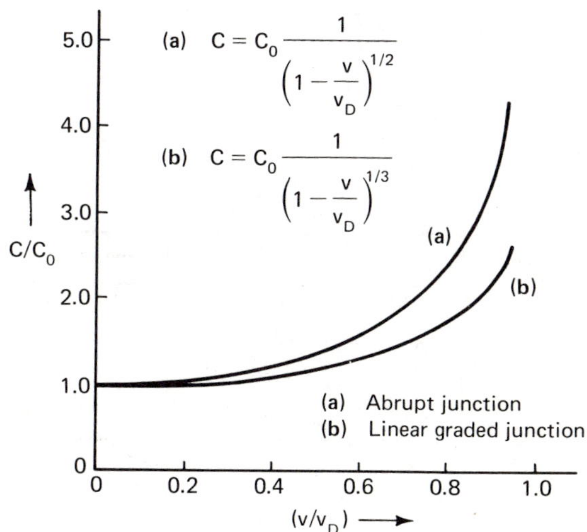

Figure 2-9: Junction Capacity vs. Voltage

Since a pure abrupt junction cannot be fabricated, the kind of junction performance actually achieved in practice will lie between the two curves.

2–3 PARAMETRIC AMPLIFIER (PARAMP) PRINCIPLES

2–3.1 A Simple Parametric Amplifier

A parametric amplifier is one that uses some regular change of an electrical parameter in a circuit to achieve electrical signal gain. The energy supplied by the operation of changing the electrical parameter goes into the amplification process. The parameter changed could be resistance, inductance, or capacitance. The case of capacitance alone will be considered here, as this could be achieved easily by means of bias voltage variation of a p-n junction diode, as illustrated in the previous section. Consider now Fig. 2-10a. A resonant circuit is oscillating at its natural frequency given by

$$F_o = \frac{1}{2\pi\sqrt{LC}}.$$

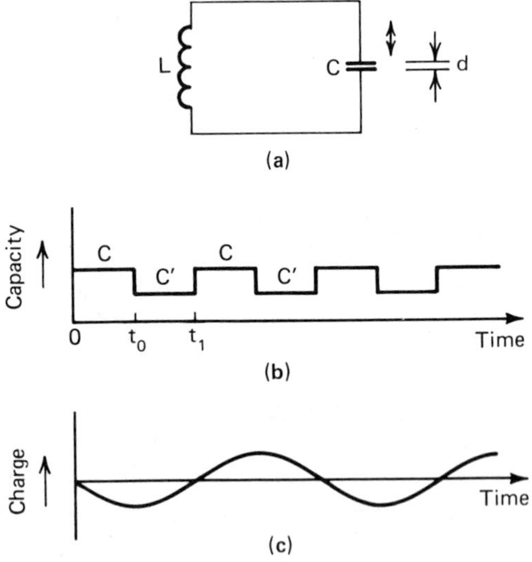

Figure 2-10: Parametric Pumping Principle

Assume that the distance d between the plates of the capacitor C can be mechanically varied. Varying the plate separation will vary the capacitance, as we have seen previously. At time t_o of Fig. 2-10b, pull the capacitor plates apart. This will decrease the capacitance C. But since the charge q is equal to the product of capacitance and stored voltage, and charge has not had time to change, the voltage across C must go up. At the next zero in the voltage cycle the capacitor can be restored to its original position. This last move will not change the stored energy, since the voltage is at zero. At the next voltage (and charge) peak the capacitor is separated again, and more energy is supplied to the circuit. It is restored again at the next voltage zero. The process of varying C at a frequency of twice the resonant frequency f_o will constantly supply energy to the circuit. (The charge stored in C and its polarity is indicated in Figure 2.10c.) It takes work to operate this "pump," and the energy expended goes into the energy of the electrical circuit. The pump must be phased properly, however, in order to maintain the oscillation. A variable capacitor amplifier operating on this principle is called a degenerate parametric amplifier.

2–3.2 The Double-Tuned PARAMP

A second form of parametric amplifier that is less sensitive to frequency and phase is given in Fig. 2-11a. A second tuned circuit called an idler is connected across the original circuit. The circuit is called a non-degenerate PARAMP, and its operation can be explained using Fig. 2-11.

Figure 2-11b traces the voltage V_1 that would appear across C if C_2 were disconnected. If C_1 were disconnected instead, the voltage across C would be V_2, as traced in Fig. 2-11c. By the superposition principle, with both tuned circuits connected, the voltage across C would be $V_1 + V_2$, as in the trace of Fig. 2-11d. If the capacitance C is pumped as shown in Fig. 2-11e, then energy will be added as in the previous degenerate case of Fig. 2-10. Note that the change in capacitance occurs at the zero value times of $(V_1 + V_2)$, which are the times of zero charge. (This follows because $q = CV = C(V_1 + V_2)$, and thus charge null must follow the voltage null.)

The condition for operation now is that $f_i = f_p - f_s$, where f_i is the idler frequency, f_s is the signal frequency, and f_p is the pump rate (frequency). Rewriting the equation as $f_p = f_i + f_s$, we see that the pump rate is identical to the sum of the signal and idler frequencies. The amount of energy gained at each zero of the pump signal is not uniform, but the energy gain is always positive. Thus, the pump phase is no longer critical for PARAMP action, and the operation is simplified because of the action of the idler circuit.

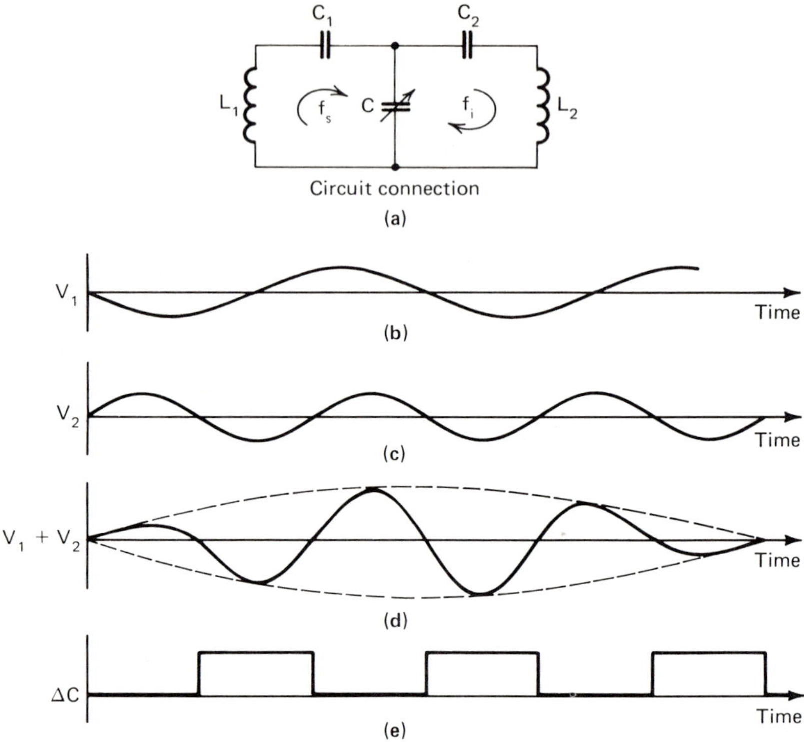

Figure 2-11: Double-Tuned Nondegenerate PARAMP

2–3.3 The Manley-Rowe Equations

A powerful tool for the analysis of parametric amplifiers is the equations derived by Manley and Rowe. They permit the development of a variety of frequency-conversion techniques and amplification mechanisms based on these idealized concepts. It first must be realized that such devices can be developed based upon parametric changes in either inductance, capacitance, or resistance. Thus, the original paper of Manley and Rowe has far–reaching consequences. We will consider here, however, only some of the methods using capacitive pumping.

Consider first the circuit of Fig. 2-12. The generators develop frequencies f_1 and f_2 respectively, and have ideal filters tuned to these frequencies. These filters have no loss—perfectly accepting the signal at

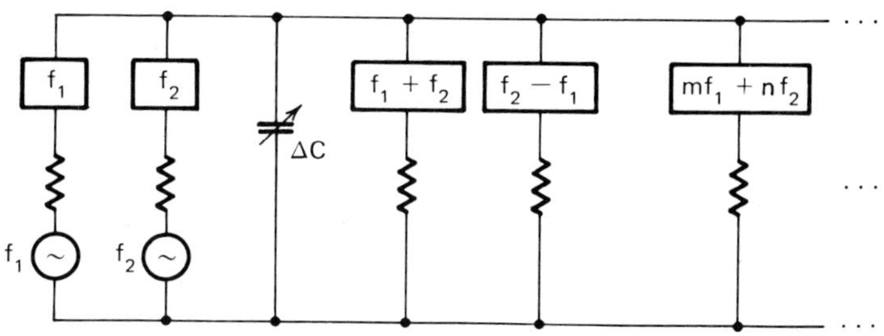

Figure 2-12: Manley-Rowe Idealized Circuit

their frequency, and rejecting completely all others. The only active sources present are the generators at frequencies f_1 and f_2.

Power will flow into and out of the ideal filters of Fig. 2-12. What is needed now is some way to predict the power flow directions so that the stability or instability of various PARAMP modes can be identified. To solve this problem, Manley and Rowe developed two systems of equations to describe the power flow. The equations will be referred to as MR–1 and MR–2. They can be written as follows:

$$\sum_{m,n} \frac{m\, P_{mn}}{mf_1 + nf_2} = 0, \text{ (MR–1)}$$

$$\sum_{m,n} \frac{n\, P_{mn}}{mf_1 + nf_2} = 0. \text{ (MR–2)}$$

The summation signs in Mr–1 and MR–2 mean summing over all possible values of m and n. As seen in Fig. 2-12, the integers m and n are given for a general filter at the end of the filter chain. Consider as a special case the filter $f_1 + f_2$. This corresponds to m = 1 and n = 1 in the general formulas $mf_1 + nf_2$. In the same fashion, the second output filter $f_2 - f_1$ corresponds to m = -1 and n = 1 in the general formula. In the use of Mr–1 and MR–2 it will be helpful to set up a table, as follows:

m	n	Frequency	Power	Role
–	–	$(mf_1 + nf_2)$	(Sign)	(Pumps, etc.)

The power flow direction will be given by the algebraic sign of P_{mn}. A positive P_{mn} value will mean flow of power into ΔC, and a negative value will mean ΔC is supplying power to the load connected to P_{mn}. In the original derivation, the ratio of f_2 to f_1 must be an irrational number, with m and n integers, as mentioned previously.

2–3.4 Analysis of Three Types of PARAMPS

Consider now the noninverting PARAMP case. Referring to Fig. 2-12, let the signal input frequency be f_1, the pump f_2, and the output $f_1 + f_2$. All other filters are removed. The analysis table will then read as follows for the input filter f_1:

m	n	$mf_1 + nf_2$	P_{mn}	Role
1	0	f_1	P_{10}	Signal Input.

Completing the table, we have

m	n	$mf_1 + nf_2$	P_{mn}	Role
1	0	f_1	P_{10}	Signal
0	1	f_2	P_{01}	Pump
1	1	$f_1 + f_2$	P_{11}	Output.

We first sum MR–1, and get

$$\sum \frac{m\,P_{mn}}{mf_1 + nf_2} = \frac{1 \cdot P_{10}}{f_1} + \frac{0 \cdot P_{01}}{f_2} + \frac{P_{11}}{f_1 + f_2} = 0,$$

or

$$\frac{P_{10}}{f_1} + \frac{P_{11}}{f_1 + f_2} = 0.$$

Similarly, we sum MR–2 and get

$$\sum \frac{n\,P_{mn}}{mf_1 + nf_2} = \frac{0 \cdot P_{10}}{f_1} + \frac{1 \cdot P_{01}}{f_2} + \frac{P_{11}}{f_1 + f_2} = 0,$$

or

$$\frac{P_{01}}{f_2} + \frac{P_{11}}{f_1 + f_2} = 0.$$

From MR–2, the pump signal at f_2 sends the power flow P_{01} into ΔC. By definition, P_{01} is positive.

Transposing the MR–2 result yields

$$\frac{P_{01}}{f_2} = \frac{-P_{11}}{f_1 + f_2}.$$

We know that $f_1 + f_2$ is positive, and thus the equation says that P_{11} must be negative. The conclusion, then, is that the power flows out of ΔC to energize the filter ($f_1 + f_2$). Returning to the result of MR–1, and transposing, we have

$$\frac{P_{10}}{f_1} = \frac{-P_{11}}{f_1 + f_2}.$$

Since P_{11} is negative, P_{10} must be positive. This equation, then, states that the power flow at the signal frequency f_1 will be into the capacitor ΔC. Thus, the input and output of the noninverter PARAMP will be stable. This form of PARAMP is also called an up-converter circuit.

For a second example, consider the PARAMP inverter case. The inverter case driver with the pump f_2 has signal input at f_1 and chooses the output at $f_2 - f_1$, thus choosing the second filter of Fig. 2-12. The analysis table for this case is written as follows:

m	n	Frequency $mf_1 + nf_2$	Power P_{mn}	Function
+1	0	f_1	P_{10}	Signal
0	1	f_2	P_{01}	Pump
−1	1	$f_2 - f_1$	$P_{-1,1}$	Output.

Summing, using MR–1, we have

$$\sum_{m,n} \frac{m\, P_{mn}}{mf_1 + nf_2} = \frac{+1 \cdot P_{10}}{1 \cdot f_1} + \frac{0 \cdot P_{01}}{1 \cdot f_2} + \frac{(-1)\, P_{-1,1}}{f_2 - f_1} = 0,$$

or

$$\frac{P_{10}}{f_1} - \frac{P_{-1,1}}{f_2 - f_1} = 0.$$

Summing, using MR–2 gives

$$\sum_{m,n} \frac{n\, P_{mn}}{mf_1 + nf_2} = \frac{0 \cdot P_{10}}{f_1} + \frac{1 \cdot P_{01}}{f_2} + \frac{P_{-1,1}}{f_2 - f_1} = 0,$$

or

$$\frac{P_{01}}{f_2} + \frac{P_{-1,1}}{f_2 - f_1} = 0.$$

Since the pump at f_2 has a positive power flow, P_{01} is positive. MR–2, then, indicates that $P_{-1,1}$ must be negative, since

$$\frac{P_{01}}{f_2} = - \frac{P_{-1,1}}{f_2 - f_1} .$$

But MR–2 states that

$$\frac{P_{01}}{f_1} = + \frac{P_{-1,1}}{f_2 - f_1} .$$

This demands that P_{01} must also be negative! Thus, the flow of power is into f_1, and a negative resistance is formed at the input. For this reason, the inverter PARAMP is called a negative resistance PARAMP. It could easily go into oscillation—or at least operate at very high gains due to this feedback of energy from ΔC back through the input filter f_1 to its generator.

The third type of PARAMP is the harmonic generator. Figure 2-13 describes this type of PARAMP. A very simple analysis table can be formed, as follows:

m	n	Frequency $mf_2 + nf_2$	P_{mn}	Comment
1	0	f_1	P_{10}	Drive Power

From MR–1,

$$\frac{m\, P_{mn}}{mf_1 + nf_2} = \frac{1 \cdot P_{10}}{1 \cdot f_1} + - - - - - = 0,$$

or

$$\frac{1 \cdot P_{10}}{1 \cdot f_1} + \frac{2 \cdot P_{20}}{2\, f_2} + \frac{3 \cdot P_{30}}{3\, f_3} + - - - - = 0.$$

From MR–2, all terms are zero, since $n = 0$. Thus, from MR–1:

$$\frac{P_{10}}{f_1} = \sum_{m=2}^{\infty} - \frac{m\, P_{mo}}{m\, f_1} ,$$

or

$$P_{10} = \sum_{m=2}^{\infty} - P_{mo}.$$

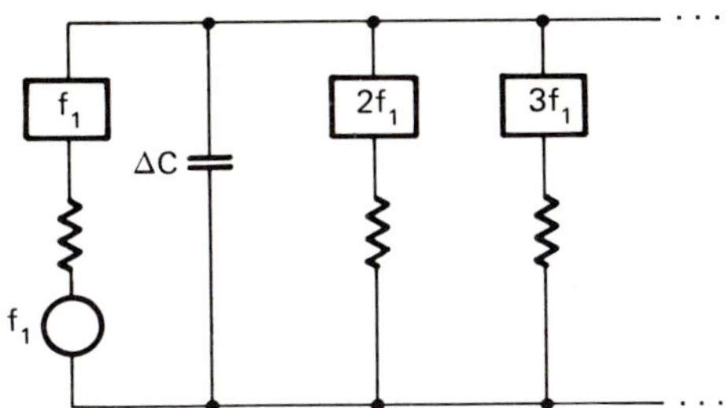

Figure 2-13: Harmonic Generator PARAMP

This result dictates that conversion from P_{10} to P_{m0} can be performed at 100% efficiency for any single value of m. It also states that when the conversion is performed, the power flow is always in the proper direction, giving maximum stability. The input power P_{10} flows from the driver generator at f_1 into the nonlinear capacitor ΔC. The power P_{m0} at harmonic m flows out of ΔC into the filter mf_1, providing power into the load following mf_1.

2–3.5 The Double-Resonant PARAMP

A Manley-Rowe type of analysis can be performed on the double-resonant PARAMP of Fig. 2-14. (Note the similarity to the double-tuned PARAMP of Fig. 2-11.)

The pumping action will be due to a variation of the capacitance at a frequency f_3 that will be the sum of the resonant frequencies of the two parallel resonant circuits. That is,

$$f_3 = f_2 + f_1,$$

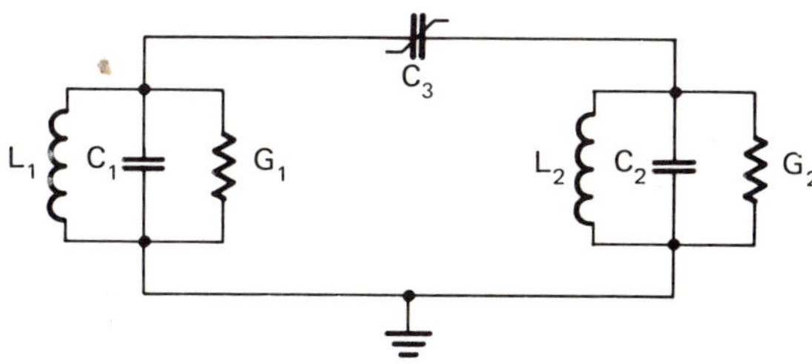

Figure 2-14: The Double-Resonant PARAMP

where

$$f_1 = \frac{1}{2\pi\sqrt{L_1 C_1}},$$

$$f_2 = \frac{1}{2\pi\sqrt{L_2 C_2}}.$$

In other words, the two "tank" circuits composed of L_1, C_1, and G_1 and L_2, C_2, and G_2 have separate resonant frequencies f_1 and f_2, respectively. They are coupled by a capacitor C_3 that has a time variation of capacitance given by

$$C_3 = C_0 \sin(2\pi f_3 t + \phi_3).$$

The time variation of C_3 is the desired pump action, and occurs at the frequency f_3.

The usual table will be constructed to investigate the stability of this PARAMP. The analysis is as follows:

m	n	Frequency $mf_1 + nf_2$	Power P_{mn}	Sign	Function
1	0	f_1	P_{10}	$-$	Signal
0	1	f_2	P_{01}	$-$	Idler
1	1	$f_1 + f_2 = f_3$	P_{11}	$+$	Pump.

Summing, using MR–1:

$$\frac{1 \cdot P_{10}}{1 \cdot f_1} + \frac{P_{11}}{(f_1 + f_2)} = 0;$$

and using MR–2:

$$\frac{1 \cdot P_{01}}{1 \cdot f_2} + \frac{P_{11}}{(f_1 + f_2)} = 0.$$

Since P_{11} is positive (i.e. the pump delivers power), MR–1 indicates that P_{10} must be negative. Similarly, MR–2 indicates that P_{01} is negative. This is true because of the positive pump power and the fact that neither m nor n is negative in value. These results appear in the sign column of the table above. This analysis indicates that power is supplied from C_3 to *both* the idler and the signal tank circuits. For this reason, it is clear that the input circuit is not stable, but regenerative (tends to oscillation). The oscillation can be controlled by adding enough conductance (G_1) in the input tank to prevent oscillation. The power flow to the idler is not a problem, as it does not receive the input signal directly. There is a small problem with developing this type of PARAMP, as it must operate as a two-terminal amplifier if amplification at f_2 is desired. In this case a microwave circulator is often used to permit separation of the input and output powers. Figure 2-15 shows the circuit connection for this type of operation. The cited problem is not serious, as it is commonly faced whenever a microwave two-terminal device is used as an amplifier. Chapter 4 outlines the same technique in the designing of a TED amplifier.

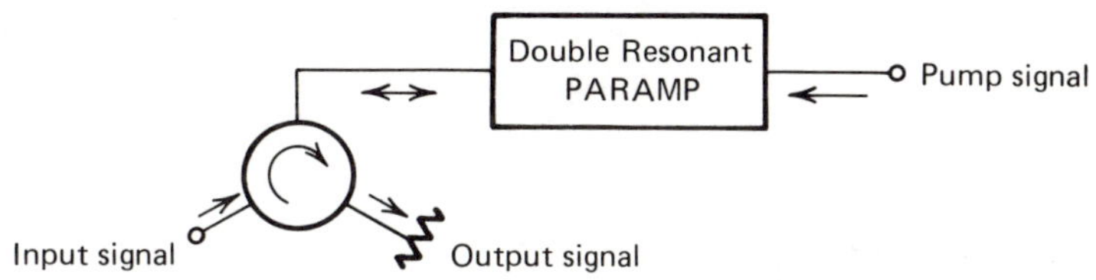

Figure 2-15: Practical Double-Resonant Circuit

2–4 EXAMPLES OF PARAMETRIC AMPLIFIERS

A basic frequency multiplier is illustrated in Fig. 2-16a. The input tuner resonates at f_1, and the output tuner resonates at $2f_1$.

In Fig. 2-16b an idler system is used. The input is tuned to f_1 and the output resonates at $3f_1$. The systems should provide relatively good efficiency, as the theoretical limit is 100% (see Harmonic Generator PARAMP, Section 2.3.4).

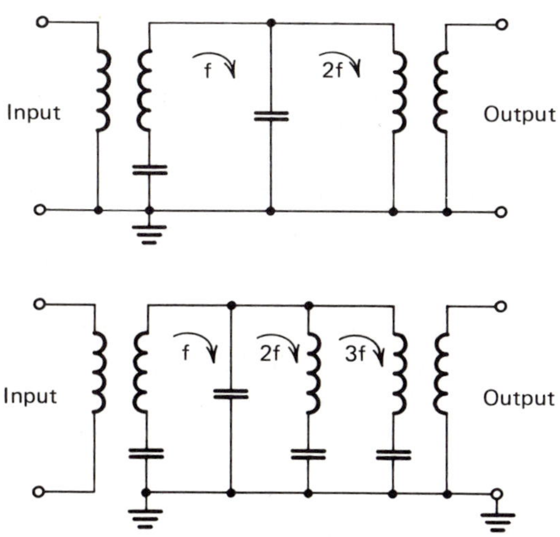

Figure 2-16: PARAMP Examples

3 Bipolar and Unipolar Microwave Transistors

3-1 INTRODUCTION

The previous chapter considered solid-state diodes and their use as parametric amplifiers. The invention of the solid-state transistor (contraction for *trans*fer res*istor*) by Shockley and his associates at the Bell Laboratories in 1948 had a strong impact on the electronic technology. As a result, transistors have to a large degree replaced vacuum tubes in low-power applications. The technology of the microwave power transistor had advanced within the previous decade. The microwave transistor is a development of the lower frequency transistor technology. To operate at microwave frequencies, the requirements on dimensions, process controls, heat sinking, and packaging become more stringent. The upper frequency limit that is possible using state-of-the-art bipolar technology is illustrated by Fig. 3-1. For comparison, a gallium arsenide field effect transistor (FET) has been added. The FET shows superior gain performance above 10 GHz, but other factors must be included. As stated in section 2–1, transistors are not competitive above 3 GHz. Clearly, they do not retain gain performance above this frequency. However, as discussed below, transistors have a number of other advantages that should be considered. This chapter will consider both the bipolar and FET technologies.

The devices considered in this chapter are generally three-terminal devices, as opposed to the two-terminal devices to be discussed in Chapters 4 and 5. The three-terminal device has the advantage that the input and output circuits are separate, providing natural isolation. The two-terminal device requires special ferrite isolation techniques to bring about the desired isolation of the input and output signal paths. These special isolation techniques are treated in Chapter 6.

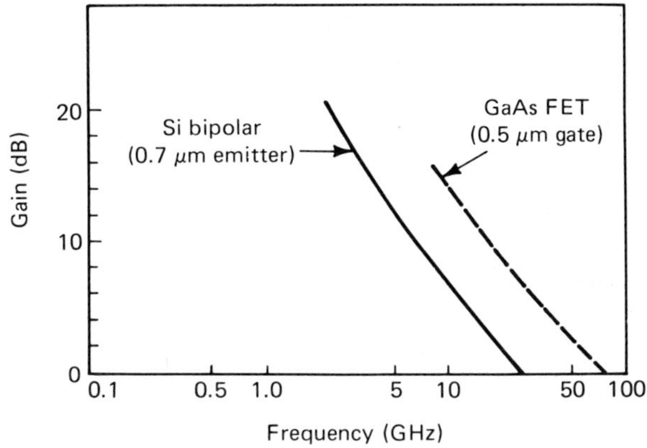

Figure 3-1: Comparison of Maximum Available Gains of Bipolar
and FET Devices

3–2 BASIC BIPOLAR TRANSISTOR PRINCIPLES

The basic transistor is a bipolar device power amplifier that uses the
nonlinear properties of doped semiconductor junction materials to achieve
the amplification. William Shockley and his co-workers at the Bell Laboratory invented the device in 1948. Figure 3-2 illustrates a p-n-p transistor, properly biased as an amplifier. The three terminals are the emitter,
base, and collector marked e, b, and c, respectively. They correspond to
the cathode, grid, and plate terminals of a conventional vacuum tube.
Figure 3-2 then corresponds to a vacuum tube amplifier operated in the
grounded-grid mode. As in the grounded-grid vacuum tube circuit case,
higher operating frequency is possible using this mode.

The bias on the input side of the transistor is forward, resulting in
the injection of electrons into the emitter p-type material and the injection

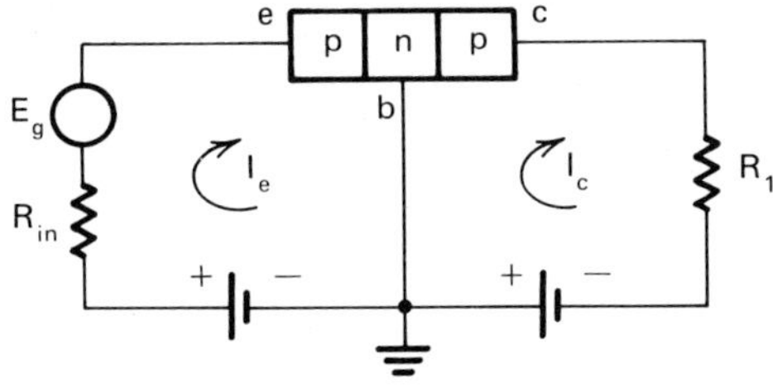

Figure 3-2: P-N-P Transistor Amplifier

of holes into the n-type material of the base. The emitter current I_e is then composed of both electron and hole currents in the semiconductor junction. Changes in the emitter current due to the signal E_g result in changes in the collector current I_e, but the net power into the load resistance R_1 will be a power gain over that in the input circuit.

3–3 THE MICROWAVE BIPOLAR DEVICE

The conventional operation of a transistor is in the Class A,* or linear, mode of the lower frequencies. At microwave frequencies, however, Class C operation is preferred. Class C operation requires biasing both the emitter-base and the collector-base junctions in the reverse mode. No current then flows until a signal is applied.

Consider now the amplifier of Fig. 3-2, but with a small reverse bias applied to the emitter-base junction. (A larger reverse bias is applied to the collector-base junction.) The applied bias levels result in a smaller depletion layer at the emitter junction than at the collector junction. With a sufficiently large rf input signal E_g, the emitter-base junction is forward-biased during the peak of the rf cycle, and electrons are injected into the base region. These injected carrier electrons transit the base by a combination of diffusing and drift flows and then are accelerated in the collector-base depletion region. The electric field in the latter depletion region is large enough, even at the minimum swing of the rf signal, to accelerate the electrons to their saturation velocity in the semiconductor material. The net result is that the flow of electrons injected into the base of the transistor at the peak of the rf cycle results in a pulse of current in the collector circuit, producing rf power in the load r_1 at the fundamental rf frequency.

Figure 3-3 gives the current-equivalent circuits for common base and common emitter microwave transistors.

Referring to the figure, the common base current gain α is given by

$$\alpha = \frac{\partial I_c}{\partial I_e}\bigg|_{V_{cb}} = \text{constant}$$

---

*Class A, B, and C operations are defined as in vacuum tube amplifier design. In Class A operation, current flows in the output of the device throughout the input rf cycle. In Class B, it flows over half of the cycle. In Class C, it flows for less than half of the rf cycle.

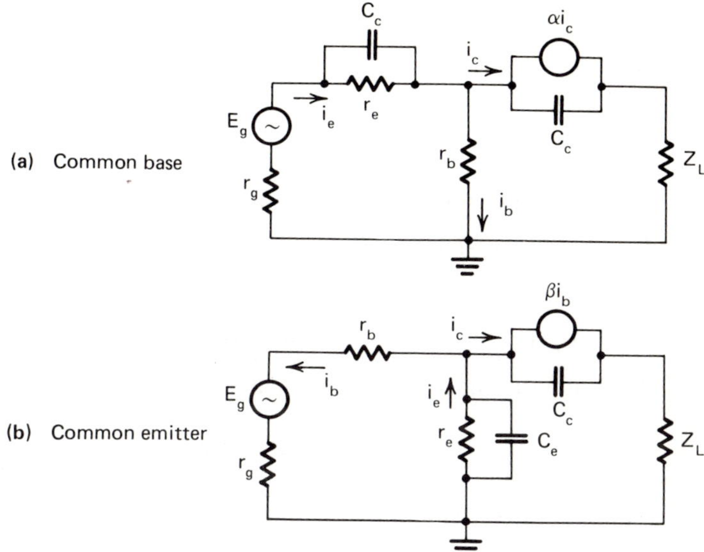

Figure 3-3: Microwave Transistor Current-Equivalent Circuits

and the common base current gain β is given by

$$\beta = \left.\frac{\partial I_c}{\partial I_b}\right|_{V_c} = \text{constant.}$$

The relationship between α and β is given by

$$\beta = \frac{1 - \alpha}{\alpha}.$$

The cutoff frequency of a charge-control type of microwave transistor is given by

$$f_T = \frac{1}{2\pi\,\tau} = \frac{\mathcal{V}}{2\pi\,L_{ec}},$$

where τ is the average time for a charge carrier moving at an average velocity $\mathcal{V}$ to flow across the emitter-collector distance L_{ec}.

An interesting upper limit for the power of a microwave transistor was derived by Johnson. The expression is

$$(P_m\,X_c)^{1/2}\,f_T = \frac{E_m\,\mathcal{V}_s}{2\pi}$$

where

$$P_m = V_m I_m$$

$$X_c = \frac{1}{\omega_T C_0}$$

$\mathcal{V}_s$ = maximum possible saturation drift velocity

C_0 = collector-base capacitance

E_m = maximum electric field

$V_m = E_m L_{min}$ = maximum allowable applied voltage

I_m = maximum current of device.

Figure 3-4 is a plot of the above equation plus a manufacturer's test results for comparison.

The equation used to plot the theoretical curve of Fig. 3-4 can also be stated as

$$(V_m \, I_m) \, f_T = P_m \, f_T = \text{constant.}$$

This formula states that, in terms of absolute limits, the maximum power for the microwave transistor must go down if the cutoff frequency f_T is increased. This seems reasonable, since f_T could be increased by reducing the emitter-collector distance L_{ec}, and this in turn would cause V_m to drop. The upper limit of f_T will be determined by the state-of-the-art for microwave transistor fabrication. For some types of devices, L is limited to 15 μm, which, for a saturation velocity of 6×10^6 cm/s, gives an f_T value of 0.38×10^9 Hz. The experimental electron drift velocity for silicon

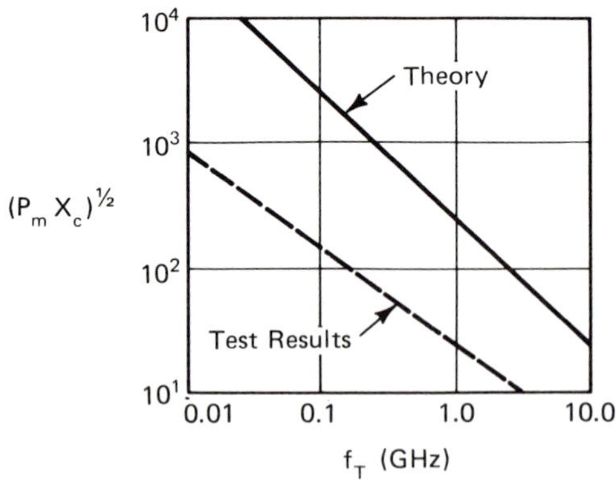

Figure 3-4: Power-Frequency Limitation of Microwave Transistors

is limited to about 10×10^6 cm/s for electric fields exceeding 1.6×10^4 v/cm. This value for the saturation velocity would give an f_T value of 0.63×10^9 Hz. Further improvement in the cutoff frequency will depend upon the increase in fabrication precision for the transistor, which sets the frequency limit for the bipolar microwave transistor.

3–4 THE MICROWAVE UNIPOLAR (FET) TRANSISTOR

After the discovery of the bipolar transistor, Shockley and his co-workers invented the field effect transistor (FET) in 1952. The FET uses a transverse electric field to modulate the conductivity of a layer of semiconductor material. The conventional transistor uses both majority and minority charge carriers in its operation. Thus, it is called a bipolar device. But the FET uses only one type of charge carrier, and for this reason it is termed a unipolar transistor.

The unipolar FET has many advantages over the bipolar junction transistor. First, it can operate up to 10 GHz in frequency. Second, it has a very high input resistance—up to several megohms. Third, it has a lower noise figure, and thus is an improved preamplifier device. Fourth, it operates at higher frequencies. Fifth, it can have both a voltage gain and a current gain.

There are basically two types of microwave FET transistors that are discussed below. One type has replaced the parametric amplifier in airborne X-Band radar systems, being less expensive and simpler to fabricate. Thus, the FET has become competitive with parametric amplifiers in noise figure performance, giving it a decided edge in the microwave solid-state amplifier field.

3–4.1 The Junction Field Effect Transistor (JFET)

The original FET proposed by Shockley was the junction FET, or JFET. Figure 3-5 illustrates an n-channel JFET. The three terminals are called the source, gate, and drain. The source terminal is similar to a cathode, and the drain is similar to a plate terminal. The gate terminal is by analogy the grid or control electrode terminal. The similarity to a vacuum tube triode is now apparent. The grid (gate) terminal is also a high impedance connection.

The schematic (Fig. 3-5b) of the n-channel JFET shows n-type material sandwiched between two heavily doped p-layers. The two p-type gates are electrically connected to form the gate terminal of the JFET.

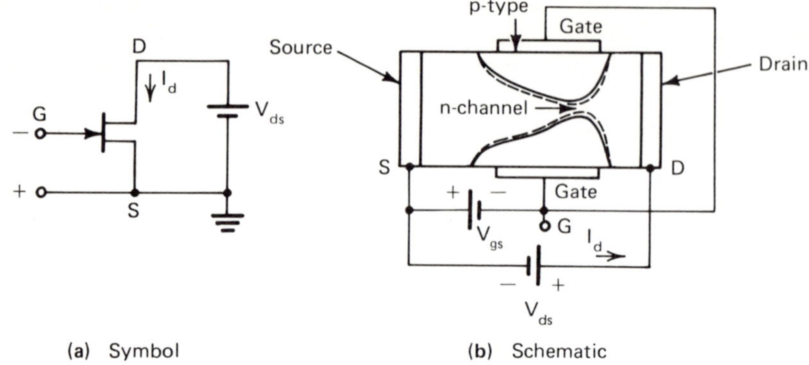

Figure 3-5: N-Channel JFET

Each end of the n-type material has a metallic contact. The source connection supplies the electrons to the device, and the drain is the contact for removing these electrons. The conventional current flow direction is indicated by I_d in Fig. 3-5b, and also in Fig. 3-5a. The symbol for the device is given in the latter. When the gate potential is varied, the charge distribution will change in the device. The current flow I_d can be cut off by applying V_{gs} large enough to cause the charge depletion regions to meet, as indicated by the dashed-line distribution in Fig. 3-5b.

The material used in the channel can be either silicon or gallium arsenide (GaAs). The channel can be either n- or p-type. A p-type channel would require reversed biased potentials and heavily dosed n-type gates. The direction of I_d would also be reversed on the circuit symbol.

3–4.2 The Schottky-Barrier Field Effect Transistor (MESFET)

The second, and more popular, type of microwave FET is the Schottky-barrier FET. It was developed by many engineers and scientists (Mead, Hooper, et al) and is termed the metal-semiconductor field effect transistor (MESFET).

Figure 3-6 illustrates the MESFET using GaAs n-type material.

The MESFET is made up of an n-type GaAs epitaxial* film desposited on a semiinsulating substrate. The gate is made up of evaporated aluminum. The source and drain contacts are Au-Ge, Au-Te, or Au-Te-Ge alloys. The n-channel layer is doped with either sulfur or tin.

---

*This film is a single-crystal substrate formed upon the substrate (semi-insulating). The word *epitaxy* is derived from the Greek επι (upon) and τελνεν (arranged).

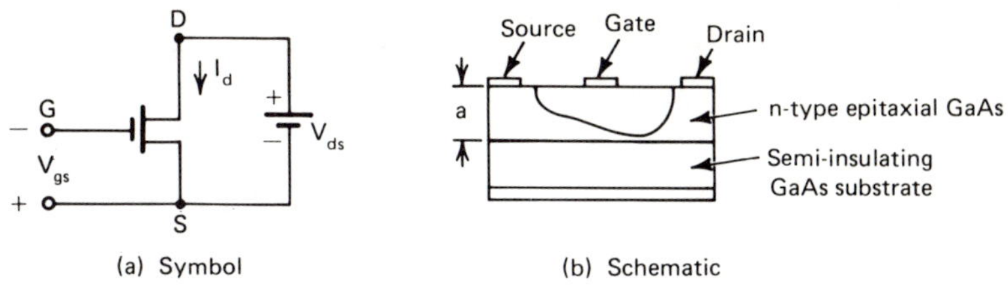

Figure 3-6: MESFET Design and Symbol

3–4.3 Principles of Operation

3–4.3.1 Some common properties

For either the n-channel JFET or the MESFET, the p-n junction between the source and gate is reverse biased. The source and drain, on the other hand, are forward biased. The majority of the carriers (electrons in the n-type material) flow from the source, through the channel between the gate electrodes, and on to the drain. The voltage drop in the material due to the majority carriers makes the gate more heavily biased as the drain electrode is approached. The charge depletion region in the channel thus expands as the drain is approached. Thus, the channel is gradually pinched off and the channel resistance increases. The result of this is that the drain current will be modulated by the gate voltage. This really amounts to a modulation of the total resistance of the channel by means of varying the depletion-layer dimensions via the gate potential. With sufficient gate potential, the channel in the vicinity of the gate can build up a very high resistance, and the drain current will be nearly constant as the drain voltage is increased.

The depletion region effect is similar to that found in the semiconductor diode of Section 2–2.2. The majority carriers are not allowed in the region due to the presence of the electric field. In other words, charge carriers are swept out of the region by the field. More precisely stated, the free charge carriers are swept out, leaving the field due to the remaining bound charges plus the field due to the bias voltage sensed by the gate. As noted above, the gate sees more bias looking toward the drain than it does looking toward the source, because of the voltage drop across the channel between source and drain.

Figure 3-7 is a plot of the drain characteristics of a typical GaAs unipolar Schottky-barrier FET, showing the variation of drain current (I_d) vs. drain voltage (V_{ds}) for fixed values of gate voltage (V_{gs}). The transconductance of the device is given by

$$g_m = \left. \frac{\partial I_d}{\partial V_{gs}} \right|_{V_{ds} = \text{constant}} \quad \text{mhos.}$$

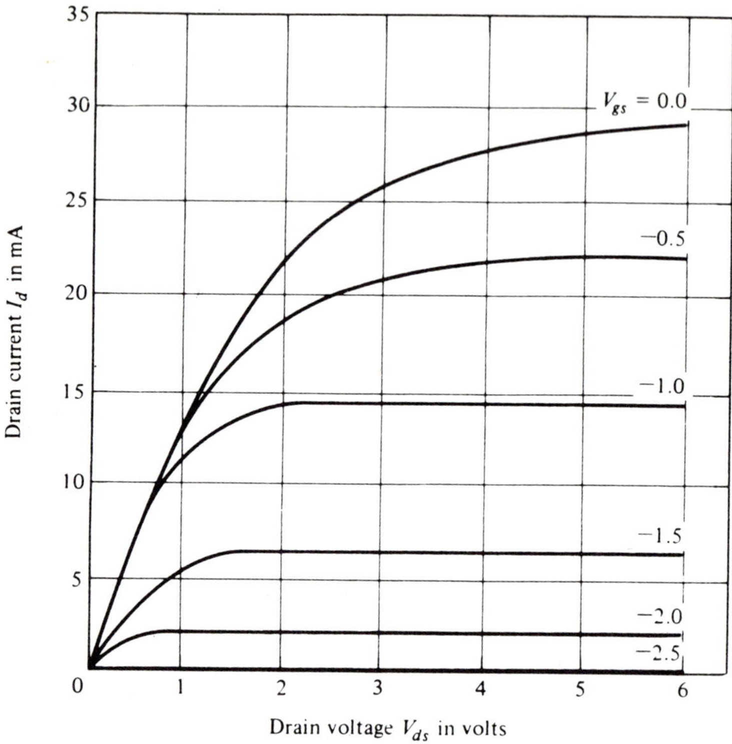

Figure 3-7: Drain Characteristics of a Typical Schottky Barrier-Gate FET *

The quantity g_m is a measure of the ability of the gate voltage to control the drain current, and thus is a measure of the amplification ability of the Schottky-barrier gate FET transistor. The voltage gain of the device is generally proportional to g_m; thus g_m is an important device parameter.

3–4.3.2 Pinch-off in the FET

Figure 3-8 illustrates the drain characteristic curve for a JFET with $V_{gs} = 0$; the region from O to P is an ohmic region, where the resistance tends to be constant. In the region from P to Q on the curve, the device is said to be "pinched off." In the region beyond Q, the JFET will experience avalanche breakdown, similar to that of a Zener diode. The point here is that pinch-off is not a zero-current condition, but one in which a form of saturation occurs. For the n-channel JFET of Fig. 3-4b, pinch-off occurs when the two depletion regions due to the upper and lower gate bias levels just touch. As the channel width approaches zero, it drops the

---

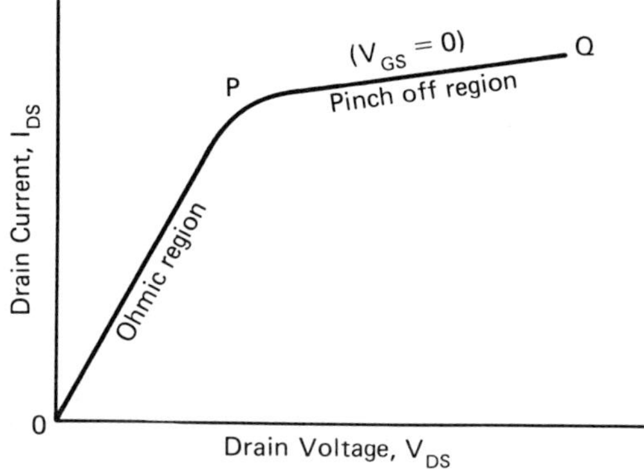

Figure 3-8: Drain Characteristic Curve for a JFET Device

current flow in the channel, which then drops the voltage gradient along the channel. This drop in voltage gradient (or local electric field) reduces the width of the depletion region, which causes the current to increase. The result is a stabilized current value in a state of equilibrium. The slope of the curve of Fig. 3-8 in the pinch-off region is what is called the *dynamic drain resistance* r_{ds}, defined as follows:

$$r_{ds} = \left. \frac{\Delta V_{ds}}{\Delta I_d} \right|_{V_{gs} = \text{constant}} \quad \text{ohms.}$$

For the case of the MESFET of Fig. 3-5b, pinch-off occurs when the depletion region beneath the gate extends down to the end of the epitaxial GaAs region, just touching the semiinsulating GaAs substrate. In either case, JFET or MESFET, the pinch-off condition occurs when enough gate voltage is applied to remove all the free charges from the channel.

Note how, in Fig. 3-7, the drain family curves for a MESFET show a pinch-off region similar to the JFET of Fig. 3-8 for gate voltages less than -1.0 volt. Normally, the breakdown region is not shown in the FET drain family curves, as the devices are seldom operated in the breakdown (Zener) region of the characteristics.

It is clear from Fig. 3-7 that operation in the pinch-off region at some reasonable drain voltage (such as four volts) will give a somewhat variable transconductance value g_m for large signal excursions. There will also be some variation in g_m from device to device. This presents some problems, as the device gain is proportional to g_m, as pointed out previously.

A simple derivation will also show that the pinch-off voltage for a MESFET can be expressed as

$$V_p = \frac{q \, a^2 N}{2\epsilon},$$

where

$$q = \text{channel free charge}$$
$$a = \text{depth of channel}$$
$$N = \text{electron concentration (electrons/m}^3)$$
$$\epsilon = \text{permittivity of channel (farads/m)}.$$

This equation shows that the pinch-off voltage is proportional to both the doping concentration and the square of the maximum channel depth.

3–4.3.3 A note on the noise figure

The intrinsic noise figure of a GaAs MESFET is given as

$$F = 2 + \gamma \left(\frac{E}{E_{Sat}} \right)^3,$$

where

$$E = \text{electric field (v/m)}$$
$$E_{Sat} = \text{saturation electric field (300 Kv/m)}$$
$$\gamma = 6.$$

At a frequency of 10 GHz, a GaAs FET has achieved a noise figure of 6.6 dB. This noise figure is superior to any other FET or bipolar transistor at that frequency. At a frequency of 5 GHz, the noise figure for GaAs FET is about 3 dB. A silicon MESFET has a noise figure given by the approximate formula

$$F = 2 + \gamma \left(\frac{E}{E_{Sat}} \right)^2,$$

where

$$\gamma = 2.3$$
$$E_{Sat} = 1500 \text{ Kv/m}.$$

3–5 BIPOLAR AND UNIPOLAR TRANSISTOR PERFORMANCE COMPARISON

A comparison can be made between bipolar and unipolar transistor performance by examining typical power amplifier performance. Figure 3-9 gives a typical family of GaAs FET linear power amplifiers and their individual bandwidths. The small signal gains vary from 20 to 37 dB, depending on the unit chosen. The type 98210 on the figure has a bandwidth covering 2 to 8 GHz, a power output of 28 dBm, and a small signal gain of 30 dB. Other higher power output amplifiers have less bandwidth. Figure 3-10 gives a family of telecommunications bipolar amplifiers for comparison. The gains are much less, and the frequencies of operation are also much lower. Power outputs, however, can extend to 20 watts (43dBm), but the drive power required is 2.5 watts. The plot of Fig. 3-10 is deceptive, however, as the 90800 series listed is actually composed of five units to cover the 570 to 960 MHz bandwidth in smaller subbands of under 100 MHz. In the widest bandwidth of the figure (series 90130) seven separate amplifiers are needed, no one of which exceeds 200 MHz in bandwidth. Ignoring the factor of cost, the GaAs FET has much better gain and frequency band coverage than the bipolar amplifier. It also operates at higher frequencies (to over 8 GHz compared to 2.7 GHz for the bipolar amplifiers).

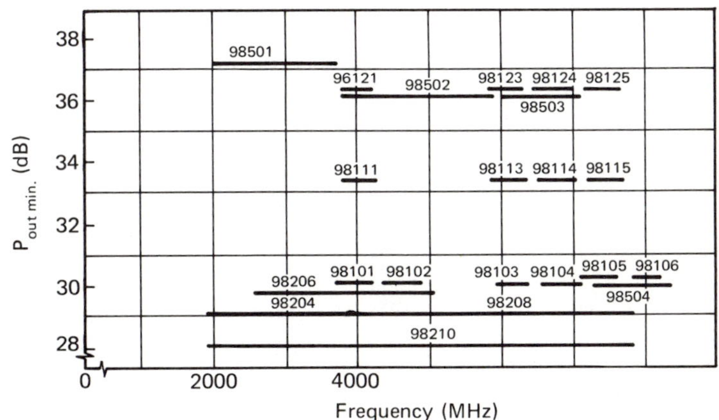

Figure 3-9: FET Power Amplifier Series

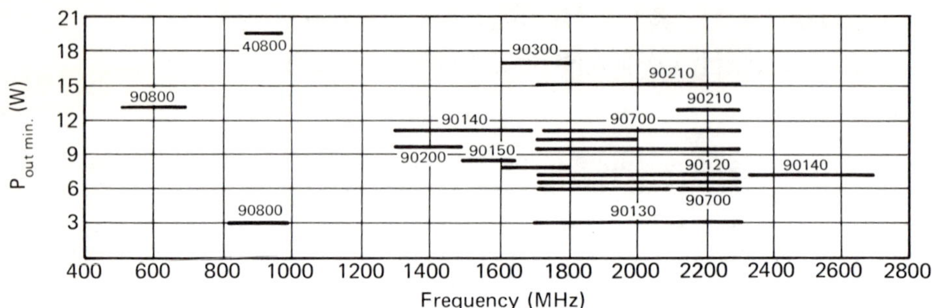

Figure 3-10: Bipolar Power Amplifier Series

The above examples of linear broadband amplifier performance serve to illustrate how the basic amplifier unit controls the performance of a multistage amplifier. It also illustrates how a manufacturer's device catalog can be misinterpreted.

4   Transferred Electron Devices (TEDs)

4–1 INTRODUCTION

The two-terminal semiconductor microwave device has seen increased use in the past decade. This device has proven to be capable of more power output at microwave frequencies than the transistor of Chapter 3. In fact, it has been proven capable of generating more cw and peak power than the best microwave power transistors. The differences between these two types of devices are very basic: (1) transistors must operate with either junctions or gates, but the TEDs employ a bulk effect without using junctions; (2) transistors are formed from elemental semiconductors, such as germanium or silicon, while TEDs are fabricated from compound semiconductors, such as gallium arsenide (GaAs), indium phosphide (InP), or cadmium telluride (CdTe); (3) transistors operate with low-energy electrons, or electrons near room temperature, while TEDs operate with "hot" electrons. Since the hot electrons have energy much greater than that of room temperature electron (0.026 eV), it is no surprise to find that TEDs can generate more power than transistors. (The greater energy translates into a higher temperature.) Thus, with the TED and the transistor operating on very different principles, the technology and theory of the two differ. In fact, very different electron circuits are needed for their operation. This chapter will explore the principles and modes of operation of the TED.

4–2 THE PROPERTY OF NEGATIVE DIFFERENTIAL RESISTANCE (NDR)

The negative differential resistance property is used in both Gunn and LSA devices. In the discussion that follows, only n-type GaAs material will be used, but the principles will also apply to other materials (InP, etc.). The principles can be explained by referring to Fig. 4-1 which is the energy–momentum relation for GaAs. In the conduction band of GaAs there are two valleys. At low electric field strengths, conduction is in the lower valley alone. At higher field strengths, electrons can make an inter–valley transfer into the bands of the upper valley. The effective mass of the electron is given by:

$$M^* = \frac{\hbar^2}{d^2E/dk^2}$$

Since d^2E/dk^2 is the rate of change of the slope of the valley curves of Fig. 4-1, it is clear that the ratio of the two effective masses will give

$$M_1^* << M_2^*$$

Thus the gradual slopes of the upper valley band will give a smaller value for d^2E/dk^2 and thus a larger effective mass M_2^*. As the applied electric field varies from 0 to 10 kv/cm the electron population in the GaAs material will gradually migrate from the lower valley to the upper valley. At a field strength of 10^4 v/cm, more than 90% of the electrons will reside in the upper valley. The mobility of the electrons is less in the upper valley due to the greater effective electron mass. An average electron mobility can be defined by:

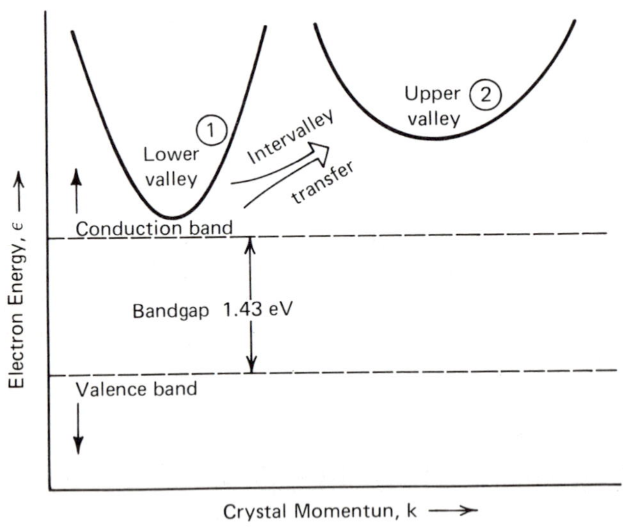

Figure 4-1: The Two Conduction Bands of Gallium Arsenide

$$\mu_{av} = \frac{n_1\mu_1 + n_2\mu_2}{n_0}$$

where

$$n_1 = \text{electrons/cm}^3 \text{ in lower valley}$$

$$n_2 = \text{electrons/cm}^3 \text{ in upper valley}$$

$$\mu_1 = \text{mobility in lower valley}$$

$$\mu_2 = \text{mobility in upper valley}$$

$$n_0 = \text{total number of electrons/cm}^3$$

Since the average electron velocity is the product of electric field and mobility,

$$\mathcal{V}_{av} = -\mu_{av}\bar{E}$$

Since, also,

$$\mu_2 << \mu_1,$$

μ_{av} will actually drop as E increases in certain ranges. Figure 4-2 shows this effect. There is an NDR region where the electron velocity will decrease as the applied field increases. Beyond the NDR region the electron

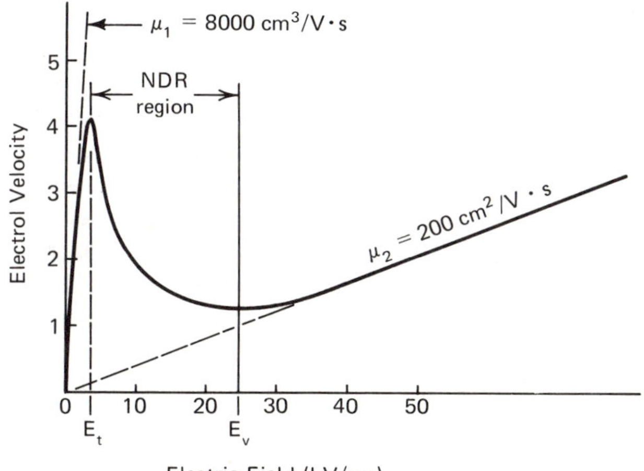

Figure 4-2: Electron Velocity vs. Field in GaAs

velocity is determined by the upper valley mobility μ_2, and the GaAs material behaves like a normal linear resistance. Since the average electrical conductivity σ is given by: $\sigma = n_0 e \mathcal{V}_{av}$ the electrical conductivity will behave as the contour of Fig. 4-2. This is the type of conductivity needed for an oscillator operating in the NDR field region.

4–3 THE NATURE OF THE GUNN MODE

The above described NDR mechanism can be used to produce microwave oscillations. To illustrate this, consider a normal dielectric. An excess charge distributed in a material will disperse (spread out) due to repelling forces according to the law

$$\rho = \rho_0 \, e^{(-\,\sigma/\epsilon)t}$$

where

$$\rho_0 = \text{initial charge density}$$

$$\sigma = \text{conductivity of material}$$

$$\epsilon = \text{dielectric constant of material}$$

In the NDR region σ is negative and small charge perturbations will grow. They will grow until their field exceeds the E_v value of the NDR region.

With this dielectric effect in mind, consider an electric field disturbance in GaAs as in Fig. 4-3. The field lies between E_T and E_V, or in the NDR region of Fig. 4-2. From this curve, the electrons in the lowest field move the fastest; therefore electrons will accumulate on the left–hand edge of the bump. The electrons at the field peak move slowest, with those right of the peak moving faster. As indicated in Fig. 4-3, a depletion region will form right of the peak, with an accumulation left of the peak. The result is the formation of a dipole charge layer in the medium that will grow into saturation and drift through the device.

The Gunn device consists of a block of n–type GaAs with ohmic contacts at either end. Figure 4-4 shows this configuration, with cathode and anode terminals. The electrons drift from cathode to anode in the device. The device is formed as a $n^+ - n - n^+$ device, with the n^+ regions serving as contacts. The terminal voltage V_a is applied at $t = 0$. This results in the field E above E_t. A dipole domain as in Fig. 4-3 forms at the cathode end and grows to maturity, and the field outside this domain

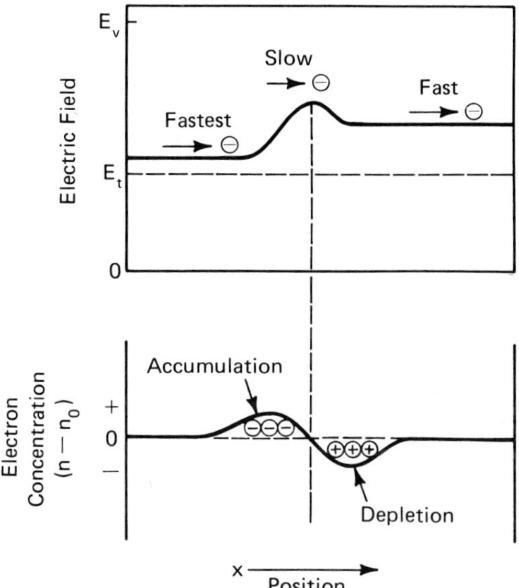

Figure 4-3: Growth of Disturbance in NDR Region

falls below E_t. This progression is shown in Fig. 4-4. The domain drifts across the sample in the time

$$t_o \cong \frac{L}{\mathcal{V}_s},$$

where

$$L = \text{sample length},$$

$$\mathcal{V}_s = \text{saturation velocity in material.}$$

The process repeats itself after the domain reaches the anode, and the field rises above the threshold again.

The Gunn oscillator frequency will then be

$$f_{Gunn} = \frac{1}{t_o} = \frac{\mathcal{V}_s}{L}$$

This Gunn oscillator will not require a cavity for resonant circuits to operate. The frequency is dictated by the transit time of the material, namely t.

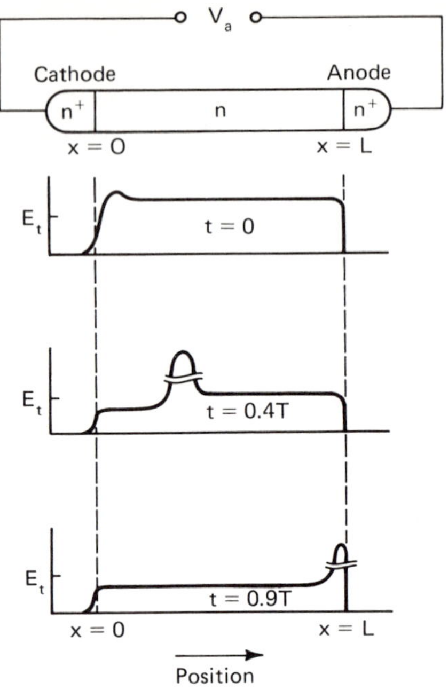

Figure 4-4: Gunn Domain Growth

4–4 DELAYED-DOMAIN (LSA) MODE

The delayed-domain, or limited space-charge (LSA), mode is more efficient than the transit-time mode. The transit-time mode, while elegantly simple, suffers from low efficiency and a frequency limitation that is determined by the thickness of the active region. The delayed-domain mode allows the Gunn device to adapt to the frequency of the external tank circuit, such as a high-Q resonant cavity. Figure 4-5 shows an equivalent circuit in which the negative differential resistance (NDR) of the Gunn device is shown as a negative conductance placed in parallel with the LC tank circuit. The conductance G_o represents the ohmic losses in the tank circuit. The circuit will oscillate if

$$| - G | \geq G_o.$$

Suppose the Gunn device is biased at some potential greater than the threshold voltage. The domain-creation phenomenon of the transit-time mode will cause several initial output pulses that serve to excite (i.e. "ring") the external tank circuit into oscillation. This action will cause

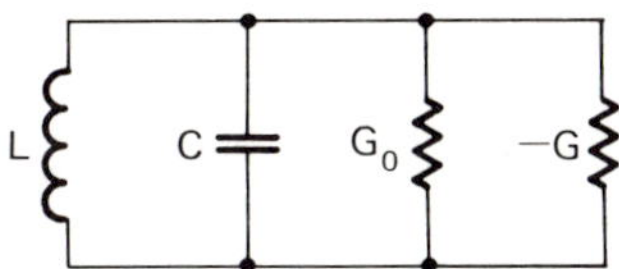

Figure 4-5: Gunn Device Equivalent Circuit

a continuous rf sine wave to build up (see Fig. 4-6) that has a frequency equal to the resonant frequency of the tank circuit. The rf voltage adds algebraically with the static dc bias such that the total bias is greater on positive peaks and less on negative peaks. The value of the static dc bias must be carefully adjusted so that the total bias drops below V_{th} on negative peaks of the rf cycle, yet will remain above the minimum sustaining potential. Whenever the total bias (i.e. the sum of the dc and rf voltages) is less than the threshold potential, the domains are quenched. If the previous domain reaches the anode while the bias is below V_{th}, then the creation of the next domain is delayed until the rf cycle brings the bias back above the threshold potential. This phenomenon causes the output current pulse period to adjust automatically to the period of the external tank circuit. The LSA mode pulse is illustrated by Figure 4-7.

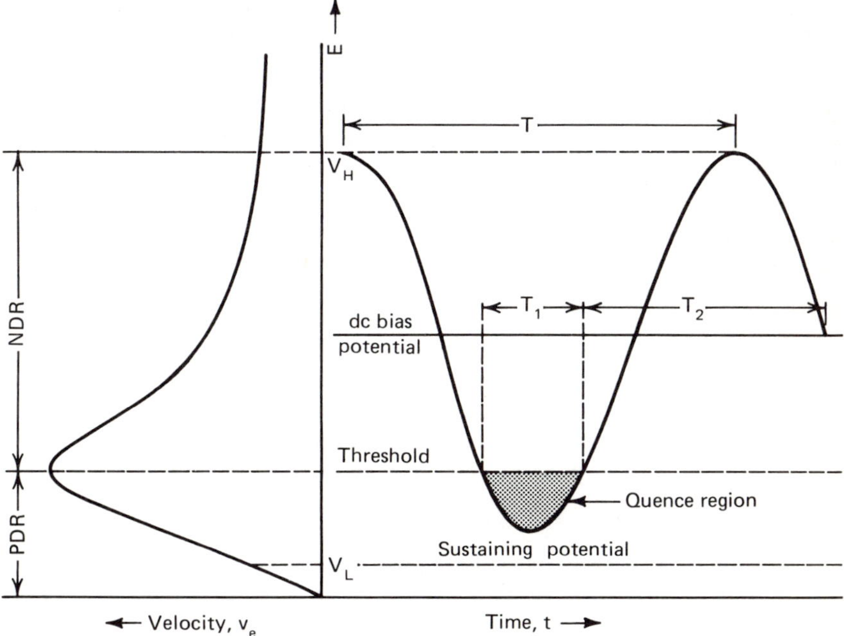

Figure 4-6: LSA Mode Oscillation Build-Up

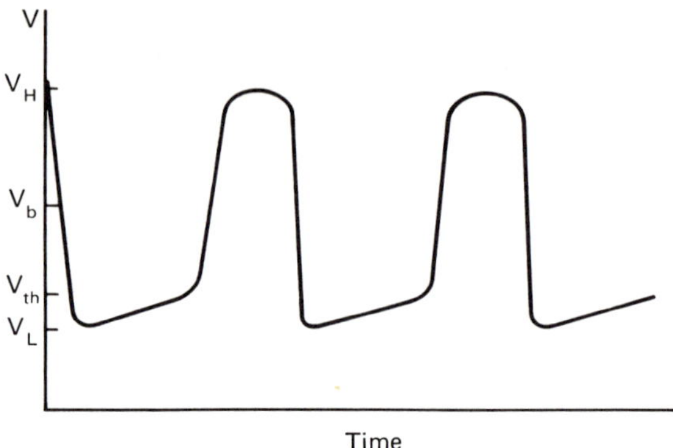

Figure 4-7: LSA Mode Pulses

We can use the frequency agility of the delayed-domain mode to frequency-modulate the device, or make it subject to automatic frequency control (AFC) operation, by manipulation of the dc bias potential.

The delayed-domain, or LSA, mode is considerably more efficient than the transit-time mode. The output power available in the transit-time mode is usually less than 1000 milliwatts, with efficiencies on the order of only 1 to 5 percent. The delayed-domain mode, on the other hand, can deliver peak powers of up to several hundred watts (with a duty cycle of 0.01 or less). The operating frequency of the transit-time mode is determined by the length of the active region and the saturated velocity of the electrons ($l/f = T = L/V_{sat}$), so it is not variable. The delayed-domain mode, however, will adjust itself to the resonant frequency of the high-Q tank circuit in which it is operated. The operating frequency can be adjusted over a range of one octave by adjusting the tank dimensions. The period is given by

$$T = \frac{1}{f} = 2\pi\sqrt{LC} + \frac{L}{R_0\,(V_b/V_{th})} \, ,$$

where

$$f = \text{frequency of operation}$$

$$L,C = \text{cavity resonance parameters}$$

$$R_0 = \text{low-field resistance (i.e., the value of R for the}$$
$$\text{case of V less than } V^{th}$$

$$V_b = \text{dc bias potential}$$

$$V_{th} = \text{threshold potential.}$$

4-5 EXAMPLES OF GUNN OSCILLATORS

The Gunn device will oscillate in the transit-time mode using only a simple resistance for the load. The efficiency in this mode, however, is only 1 to 5 percent, so relatively large amounts of dc power are required to generate small amounts of rf power.

If the Gunn device is placed inside a resonant cavity, and the device is biased for the delayed-domain mode, then better efficiency will be obtained along with flexibility of the operating frequency.

Figures 4-8 and 4-9 show two methods for mounting a Gunn device inside a resonant cavity. Figure 4-8 shows a cutaway view of a coaxial cavity. The cavity is one-half wavelength long, with the base of the Gunn device placed at the one-eighth wavelength point. A conductive "dowel" supports the Gunn device and connects it to the ends of the cavity; the dowel is also the center conductor of the coaxial cavity.

A tuning screw is used to vary the operating frequency of the device. It effectively changes the dimensions of the cavity and is capable of fine-tuning the operating frequency over a small range.

The oscillations on the inside of the cavity are coupled to the outside world through a short coupling loop that is placed parallel to the dowel center conductor. The load impedance of the Gunn device is determined by the position of the coupling loop and is adjusted for the best compromise between the stability of the operating frequency and the maximum output power.

The coaxial cavity, while simple, suffers from a few basic problems. It is a low-Q tank and is thus more sensitive to factors such as temperature

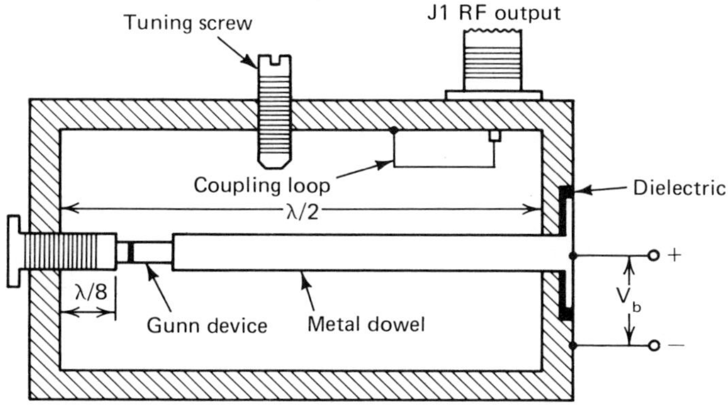

Figure 4-8: Cutaway View of Coaxial Cavity

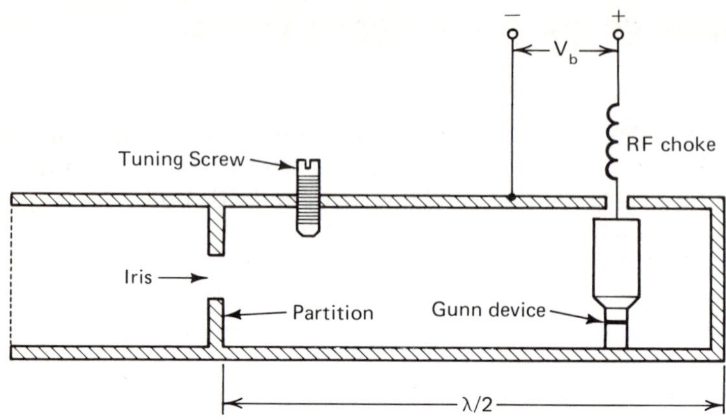

Figure 4-9: Rectangular Waveguide Cavity

and load impedance variations. The Gunn device in a coaxial cavity may also tend to oscillate on a harmonic of the tank frequency.

A rectangular waveguide (Fig. 4-9) can also be used as a tuned cavity if one end is blocked off and the Gunn device is placed at the one-eighth wavelength point. The required dc bias is provided to the Gunn device through an rf choke that is designed for microwave operation.

The dimensions of the cavity are determined by the placement of a partition. Energy from the cavity is coupled into the waveguide transmission line through an opening called an iris. The size of this iris is a trade-off between maximum output power and a sensitivity to changes in the load and internal impedances of the Gunn device.

5 Avalanche and Barrier Injection Devices

5–1 INTRODUCTION

The two-terminal semiconductor devices (Gunn oscillators) of Chapter 4 operate simply by applying a dc voltage to a bulk semiconductor. The devices have no p-n junctions. The frequency is determined by the load and the natural frequency of the circuit. The devices described in this chapter rely on the effect of voltage breakdown across a reverse biased p-n junction. This breakdown produces a supply of holes and electrons to produce the oscillations. The so-called avalanche diode oscillator uses carrier impact ionization and a drift in the high field region of a semiconductor junction to produce a negative resistance for use at microwave frequencies. The result is a negative-resistance device that can be used to produce strong microwave oscillations. The general device was first proposed by Read when he analyzed the negative resistance properties of an idealized complex diode (n^+-p-i-p^+ type).* The result of the latter was the identification of two distinct avalanche oscillator modes, the IMPATT and TRAPATT modes. IMPATT stands for <u>IMP</u>act ionization <u>A</u>valanche <u>T</u>ransit-<u>T</u>ime operation and TRAPATT stands for <u>TRA</u>pped <u>P</u>lasma <u>A</u>valanche <u>T</u>riggered <u>T</u>ransit operation. These two basic modes of operation are discussed below.

Another type of active microwave device is the BARITT diode. This acronym stands for <u>BAR</u>rier-<u>I</u>njected <u>T</u>ransit-<u>T</u>ime device. It has a long drift region and is thus similiar to the IMPATT diode. The carrier-injection mechanism differs from that of the IMPATT as noted in the discussion that follows.

---

*The superscript plus sign refers to very high doping, and the small i refers to intrinsic, or naturally occurring, semiconductor material.

5–2 IMPATT DEVICE PRINCIPLES

The IMPATT diode was first proposed in 1958 by W.T. Read of Bell Laboratories. Read's suggestion was that the phase delay in a p-n junction diode between an applied rf voltage and an avalanching current could be used for negative-resistance operation at microwave frequencies. Read's model diode would have carriers drifting through a depletion region, causing the negative resistance. Fabrication difficulties prevented the construction of a working Read diode until the mid-60's. In 1965, however, R. L. Johnson of Bell Labs verified the validity of Read's model when he generated approximately 80 milliwatts of rf power at 12 GHz from a silicon p-n junction diode. Read's diode depends upon impact avalanche and transit-time phenomena, so was given the acronym IMPATT. It has now been recongnized that Read's structure is just one of several that will result in IMPATT operation.

5–2.1 Avalanche Phenomena

Figure 5-1 shows the I-vs-V curve for a p-n junction diode. For present purposes, consider operation in the reverse bias region, i.e. the region in which V is less than zero. There is a critical breakdown voltage

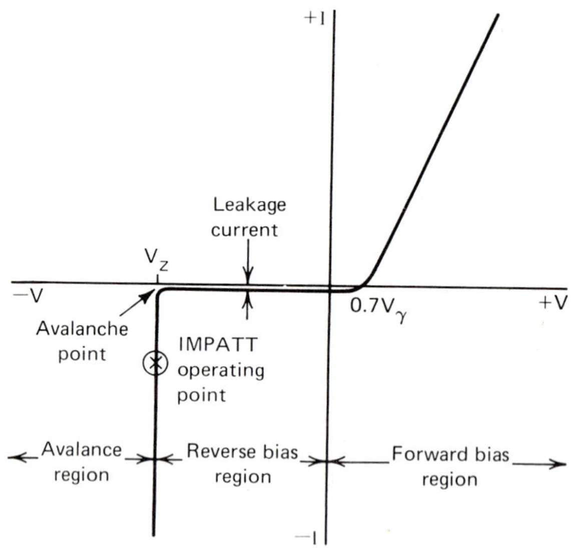

Figure 5-1: P-N Junction Volt-Ampere Characteristics

V_z in the reverse bias region. At reverse potentials less than this value, the current through the p-n junction is very small; in fact, it is a tiny leakage current. But the current suddenly increases when the voltage exceeds V_z: the junction is then operating in avalanche. The increased current is due to secondary emission, or avalanche multiplication, in which electrons of the leakage current have a high probability of being multiplied. The result is a very rapid increase in reverse current. In ordinary signal or rectifier diodes the avalanche phenomenon can be destructive. Certain types of diodes, however, are able to control the avalanche process by using properly doped semiconductor material. Zener diodes and controlled avalanche rectifiers are in this category.

Consider the p^+nn^+ IMPATT diode shown in Fig. 5-2. The p^+n junction which will be discussed is on the left side of the structure. Note that the right hand junction is an n-n^+ structure. The n^+ region forms a contact of low resistivity for the electrode, preventing metallic ion migration (much as in the Gunn structure) into the active region.

The center n region is the active zone and must be doped such that it is fully depleted (or punched–through) at breakdown. This is to insure that a very small electrical field will cause velocity saturation of the electrons. The transit time of the IMPATT is defined as $T = L/V_{sat}$. The electrons generated in the avalanche zone of the IMPATT diode will flow into the drift zone of the n-region. It takes very little added voltage to cause a large increase in current in this mode.

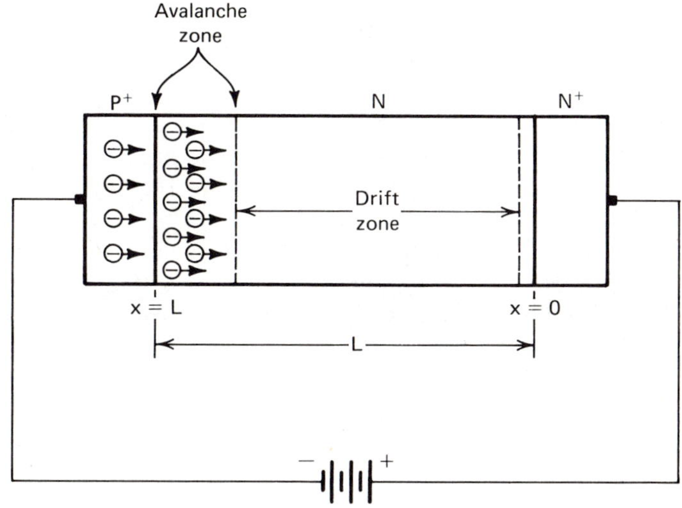

Figure 5-2: P^+-N-N^+ IMPATT Diode

5–2.2 IMPATT Oscillation

The Read diode, which was the theoretical model for the IMPATT diode, is more complex than the simple p-n junction of Fig. 5-2. However, the volt-ampere characteristics of the two will be similar so that Fig. 5-1 can represent either diode type.

The experimental work of Johnson mentioned above was performed using a p-n junction in the avalanche mode. At the time, the basic mechanisms giving rise to the negative resistance were felt to be the same as in a Read diode, even though the devices differ in construction. Now an abrupt junction will have a field distribution such that the avalanche will be confined to a narrow region, with only one carrier participating in the drift current throughout the diode. In addition, Misawa showed in 1966 that the abrupt junction has impedance characteristics approximating that of the Read diode. As a result, the abrupt p-n junction in its avalanche mode is classed with the Read diode as an IMPATT device.

Figure 5-3a illustrates the structure of the Read diode. In this case it is formed as an n^+-p-i-p^+ diode. The p-region is narrow and lightly doped so that at low levels of reverse bias the voltage appears essentially across the n^+-p junction. As the bias is increased, the excess carriers in the p-region are depleted and "punch-through" occurs (i.e., the field now extends to the i-p^+ junction). The bias is increased further until the voltage across the i-region drives the carriers into saturation current and avalanche via impact occurs in the narrow p-region. Under these conditions the current in the i-region is primarily holes drifting at the saturation velocity. (The generated electrons due to avalanche are quickly swept into the n^+ region.) The bias will now be as indicated in Fig. 5-1.

Consider now an ac voltage of Fig. 5-3c applied to the Read diode under the above bias conditions. The carrier space charge will give the injected current of Fig. 5-3d, which lays the applied ac voltage by 90 degrees. It can be concluded from this that the IMPATT is a negative resistance device.

The avalanche current pulse drifts through the i-region at velocity V_s until it is collected at the output of the n region. The drift time, then, is proportional to the length of the drift region and inversely proportional to the saturation velocity:

$$T = \frac{L}{V_s},$$

where

T = drift time in seconds

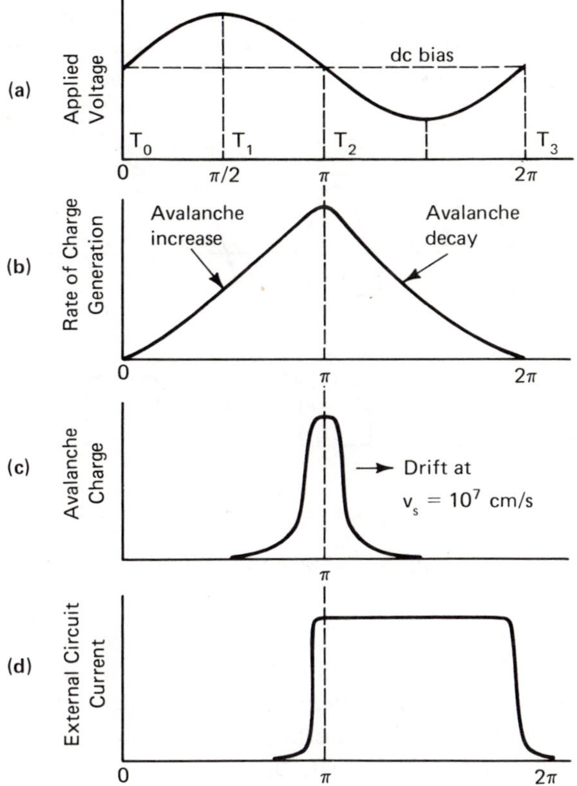

Figure 5-3: Read Diode Mechanisms

$$L = \text{path length in centimeters}$$

$$V_s = \text{electron saturation velocity } (\simeq 10^7 \text{ cm/s}).$$

The pulse current in the external tank circuit (Fig. 5-4) is square-shaped and represents a current lag over applied voltage of nearly 180 degrees. Two factors combine to cause the positive external current during the negative excursions of the rf waveform: the time delay of the avalanche process and the drift time of the avalanche charge. Instead of absorbing energy, in the manner of a positive, or ohmic, resistance, the IMPATT offers a negative resistance.

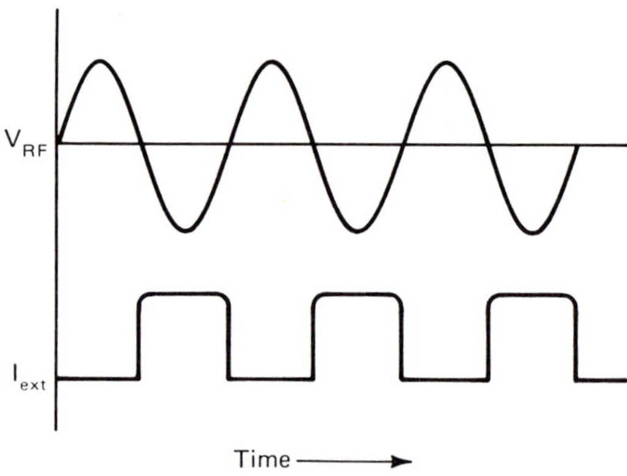

Figure 5-4: IMPATT Terminal Voltage and Current

The IMPATT device which was shown previously is known as a single-drift device. But an avalanching p-n junction produces both kinds of charge carriers, i.e. holes and electrons. The single-drift IMPATT uses only the electrons, and returns the holes to the cathode p-region. This fact limits the efficiency of the single-drift devices to less than 15%.

5–2.3 Double-Drift IMPATTs

Greater efficiency is obtained through the use of a double-drift IM-PATT device, such as that shown in Fig. 5-5. This structure is a p^+-p-n-n^+ in which the avalanche region brackets the p-n junction. The p^+ zone serves as an ohmic contact for hole charge carriers, while the n^+ region serves the same purpose for electrons. The output efficiency is increased over that of the single-drift variety because the holes drift across the p-zone very nearly in phase with electrons drifting across the n-zone.

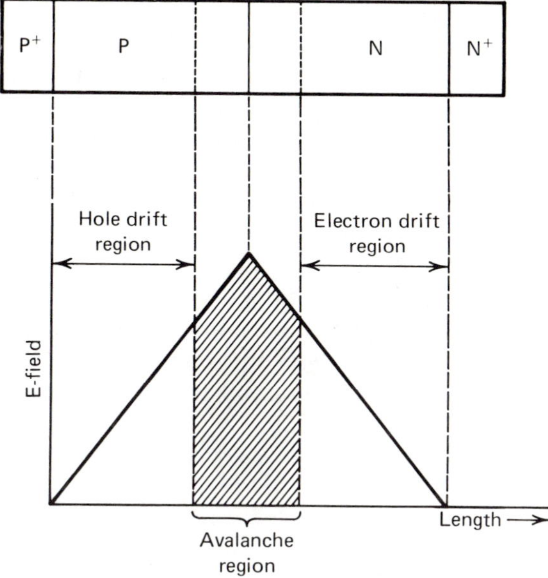

Figure 5-5: Double-Drift IMPATT Device

5–3 IMPATT APPLICATIONS

The previous discussion has demonstrated that the IMPATT device will function as an oscillator at microwave frequencies. If an IMPATT is placed inside of a high-Q resonant cavity and biased with a dc potential slightly above the avalanche potential, then noise pulses will "ring" the cavity to produce the rf sine wave that actually drives the junction into the IMPATT mode of oscillation. IMPATT operation occurs because the voltage of the ringing waveform (an rf signal) algebraically adds with the static dc bias, causing the junction to go into the avalanche mode on peaks of the rf cycle. If the device is correctly biased, then the junction will be in the avalanche condition for most of the positive half of the rf sine wave excursion.

Although the IMPATT device is an oscillator that is capable of producing substantial peak pulse powers at microwave frequencies, it is not universally applied because it is a noisy source. (Avalanching is a noisy process.) For this reason, one does not ordinarily see IMPATTs as receiver-local oscillators.

IMPATTs are used primarily at frequencies above 3 or 4 GHz, with frequencies up to 100 GHz having been obtained. Many high-power IM-

PATTs require operating potentials between 75 and 150 volts dc, a fact seen as a disadvantage by some. Also, IMPATTs are usually operated from constant current supplies, which is a potential disadvantage. The efficiencies obtained from IMPATTs range from 12 to 15 percent for single-drift devices and 20 to 30 percent for double-drift devices made of GaAs material.

The applications of the IMPATT are not limited to oscillator service. There is one report of IMPATTs being used as frequency multipliers. Many IMPATTs are used as amplifiers. In fact, it has been claimed that most IMPATTs are used as amplifiers, not as oscillators. IMPATT amplifiers have only one port and so must be coupled to a circulator in order to isolate input and output ports of the amplifier (see Fig. 5-6). This type of amplifier is called a reflection amplifier.

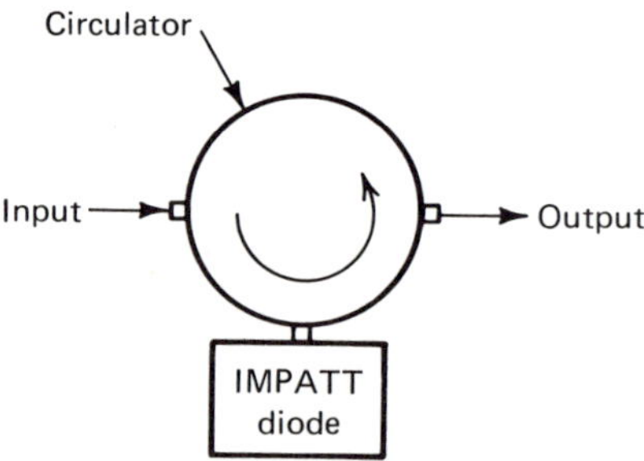

Figure 5-6: IMPATT Amplifier

5–4 TRAPATT DIODES

5–4.1 Discovery of the Mode

IMPATT diodes are generally limited to operation at frequencies above 3 or 4 GHz. The problem of lower operating frequencies (e.g. 0.9 to 3.0 GHz) is one of finding a method for stretching the duration of the transit time. Until 1967, it had proven difficult to use solid-state devices to generate any significant amount of power in the 1 GHz region. In 1967, however, engineers working for RCA succeeded in exciting the IMPATT-like device into a different mode of operation. One set of trials produced pulse powers of 425 watts at an efficiency of 25 percent. Further work with this new mode yielded efficiencies up to 60 percent, and later work produced efficiencies as high as 75 percent. Tuned-tank circuits developed at RCA in that era permitted a tuning range that was continuous over 0.9 to 1.5 GHz.

It appeared that the problem of increasing the transit time had been solved, but the mechanism was not understood. At the time the basic work on the TRAPATT device was going on, there was no good theory that explained the observed behavior. Workers at RCA dubbed the new mode the anomalous mode, perhaps reflecting the fact that they had no theory of its operation.

At least two different theories were advanced to explain the behavior of the anomalous mode. Bell Laboratories advanced the theory that the high efficiency and lowered frequency of operation were caused by the fact that a plasma was created in the device between sweeps of the IMPATT mode of operation. The theory held that the trapped plasma shielded the charge carriers from the external voltage field, causing them to drift out of the plasma at low velocity. This theory led to the acronym by which the device is now known: TRAPATT (trapped plasma avalanche triggered transit).

Cornell University physicists offered a somewhat different explanation of the high efficiencies observed. The Cornell theory maintained that avalanche resonance pumping was the responsible mechanism and advanced the acronym ARP. The Bell Labs view seems to have prevailed. The difficulty in determining the proper theory of operation caused a two-year delay between the first observations of the TRAPATT mode and the explanation of how it worked. Part of the problem is that TRAPATT operation is not amenable to small-signal analysis, so the correct theory had to be worked out somewhat more laboriously than would have otherwise been possible.

It has been demonstrated that ordinary p-n junction diodes (silicon) can be made to oscillate in the TRAPATT mode. It is, however, rather tricky to adjust such circuits, so they do not find much application. Most commercial TRAPATT devices use the p^+-n-n^+ structure of the single-drift IMPATT. A typical TRAPATT device is shown in Fig. 5-7a.

5–4.2 TRAPATT Oscillators

The structure of the TRAPATT device is very similar to that of the IMPATT. In fact, some TRAPATT devices will oscillate in either the TRAPATT or the IMPATT modes, depending upon the bias and other circuit conditions. Numerous TRAPATT oscillators actually start out in the IMPATT mode for a few nanoseconds after turn-on and then convert to the TRAPATT mode when certain circuit conditions are satisfied. In order to effect the switchover from the IMPATT to the TRAPATT mode,

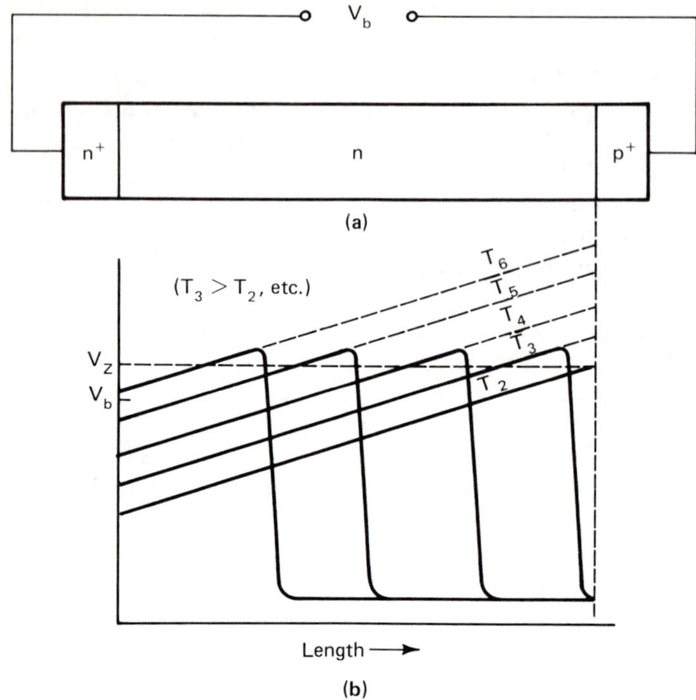

Figure 5-7: TRAPATT Device and Avalanche Shock Front

the device must be driven with a hard current pulse. Since the rise time
of this pulse must be very short, it is usual to use the IMPATT mode to
generate such pulses. (It is very difficult to obtain the rise times needed
with external circuitry.)

What is a trapped plasma, and how does it behave? A plasma exists
whenever a large density of both positive and negative charge carriers
(holes and electrons) are present. If the electric field in that region is very
low, then the plasma is said to be trapped, i.e. it takes a long time to
sweep carriers out of the region under the influence of the electric field.
The carrier velocity is considerably lower than the saturation velocity.

Figure 5-7 shows the dynamic field distribution of the n-region. The
device is biased to just punch through, i.e. the depletion zone reaches
through the entire length of the n-region but is biased at a point less than
the avalanche voltage V_z. The slope of the electric field (Fig. 5-7b) is
dependent upon the charge density. Because it is operated in the punch-
through region, there will be no free charge carriers present.

If the TRAPATT diode is excited with a large, fast-risetime current
pulse I_0, then a point in the constant bias field V_b will be observed to
move as an avalanche shock front with a velocity

$$V_x = -\frac{I_0}{eN_d}.$$

The velocity of this shock front V_x will be observed to move faster than the saturated velocity of the holes and electrons, a phenomenon that is much like the behavior of water waves at the beach striking the shore with an angle other than 90 degrees. Typical times for the avalanche shock front to traverse the n-region are around 100 picoseconds.

Notice what happens to the terminal voltage in Fig. 5-7b. Shortly after the initiation of each shock front, the terminal voltage drops from a very large value down to a very small value. The time of fall of this drop is very brief, so the TRAPATT operates as a very fast, low-impedance electronic switch. If the diode is placed at one end of a half-wavelength transmission line (see Fig. 5-8), then this phenomenon will result in a pulse being applied to the transmission line at L = 0. The current pulse has a fast risetime and a nearly square waveshape, so it is rich in harmonics. (See Fig. 5-9.) The harmonics are taken out by a low-pass filter so that the fundamental signal could be applied to the load Z_l. The value of the current pulse will be $I = V_b/Z_0$, where V_b is the applied bias potential and Z_0 is the characteristic impedance of the transmission line.

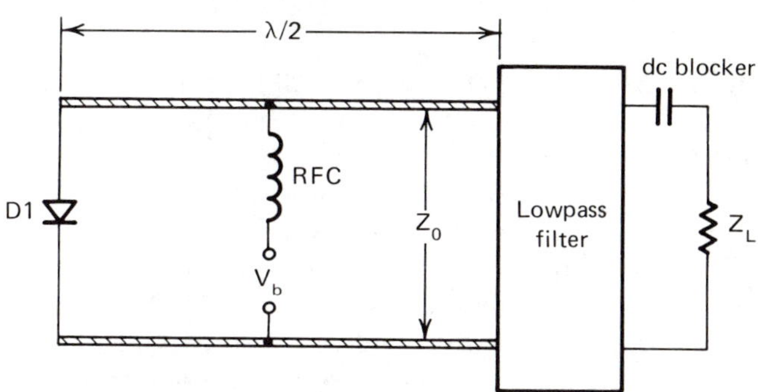

Figure 5-8: TRAPATT in Transmission Line

The description above said that a current pulse had to be applied to the TRAPATT diode before TRAPATT mode could be realized. This pulse could easily be an IMPATT pulse when the device is operating in the IMPATT mode. But the TRAPATT mode will be observed to build up in a nearly exponential manner until it becomes self-sustaining. The foregoing discussion does not explain how the TRAPATT mode could become self-oscillatory. For this type of operation, reliance must be placed upon the actions of the low-pass filter at the end of the resonant half-wavelength transmission line. The filter will transmit energy at the

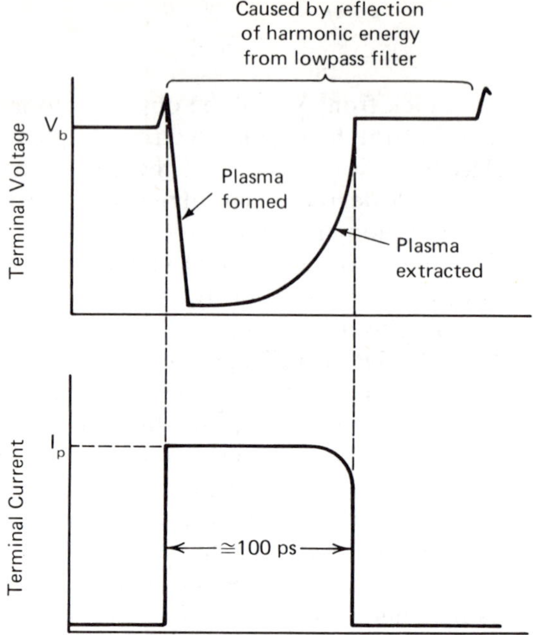

Figure 5-9: TRAPATT Drive Levels

fundamental TRAPATT frequency, but will reflect energy at the harmonics of the fundamental frequency. These harmonics are reflected with phase reversal (i.e. reflection coefficient -1), so they will initiate another avalanche shock front. The small rise in the terminal voltage shown in Fig. 5-9 is caused by this returning reflection. The return pulse will then cause the TRAPATT mode oscillations to be continuous. In the typical TRAPATT oscillator, the device will begin in IMPATT operation. The IMPATT mode oscillations will build up in an exponential manner until the current becomes large enough to trigger a shock front transit (hence the use of triggered transit in the device acronym).

5–5 BARITT DEVICES

5–5.1 Device Principles

Consider the p^+-n-p^+ structure shown in Fig. 5-10a. This device is a pair of abrupt junctions set back to back. One of these junctions will be slightly forward biased, while the other junction is slightly reverse

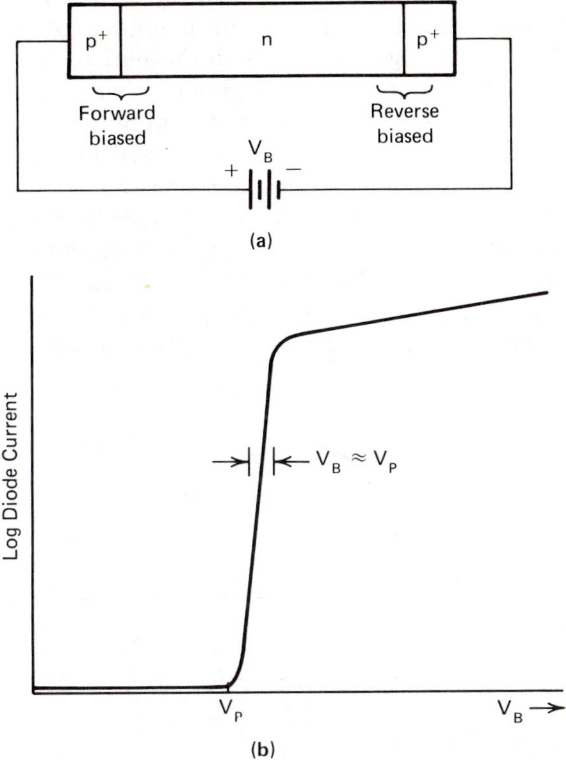

Figure 5-10: BARITT Device and Characteristics

biased. The flow of current when the bias voltage is less than the punch-through voltage will be limited by the ordinary leakage current of the reverse biased junction. If the bias is increased to the point where the device is operated in the punch-through mode, then the depletion region exists across the entire n-region until it reaches the forward biased junction. This will cause all of the carriers (holes) at the forward biased junction to be swept across the n-region, in turn causing the current to increase rapidly (see Fig. 5-10b). This current can be used in a microwave oscillator provided that (1) the field is large enough to make the holes drift across the n-region at the saturated velocity (10^7 cm/s), and (2) the voltage applied is kept below the point at which avalanching will occur. Devices that use this phenomenon are called BARITT (Barrier Injection Transit Time) oscillators.

Suppose that a BARITT device is biased with a dc potential close to the potential required for punch-through operation. Suppose further that this device is operated parallel with a resonant tank circuit (as was done in the previous Gunn device discussion). Noise pulses will ring the tank circuit and cause an rf alternating potential to appear across the diode, which will add algebraically with the bias potential. (See Fig. 5-11a.) When the total bias goes over the punch-through potential on positive excursions of the rf waveform, a sharp current pulse is injected. (See Fig. 5-10b and Fig. 5-11b.) During the period when the injected current is peaking, the terminal current (from the dc bias) is added to it, causing a reversal of the current direction for that period. (See Fig. 5-11c.) If the power, i.e. the product of Fig. 5-11a and 5-11b, is plotted, the power-vs-time waveform of Fig. 5-11d will be obtained. Notice that for substantial periods during the cyclic excursion, the power is negative, meaning that the device is oscillating and will deliver energy to the external tank circuit.

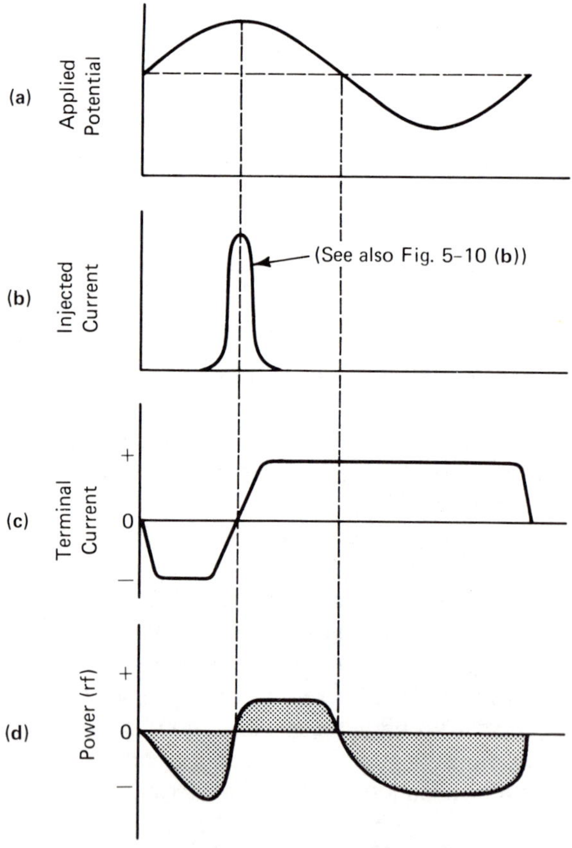

Figure 5-11: BARITT Parameters

There is only a brief period in which the terminal current and terminal voltage are both positive (T_2–T_3), and this will limit the efficiency of the BARITT oscillator.

5–5.2 Examples of Applications

The BARITT device is a low-power microwave energy source and is, in many ways, considered superior to the Gunn device for service such as microwave receiver local oscillators, doppler intrusion alarms, and certain shoplifting protection devices.

6   Solid-State Microwave Amplifiers

6–1 INTRODUCTION

In the discussion of two- and three-terminal devices in the introduction to Chapter 3, mention was made of the need for special circuit design to permit the use of a two-terminal device as an amplifier. Chapters 4 and 5 considered two types of semiconductor diodes (TEDs and Avalanche/Barrier Injection Devices), some of which can be used as amplifiers. In general, these diodes, when considered as negative resistance elements, are capable of power amplification. In fact, any negative resistance element can bring about power gain. The problem is that a two-terminal device does not separate the input from the output circuit. The devices of Chapter 3, bipolar and unipolar, are three-terminal devices and thus have a degree of isolation between the input and output circuits.

This chapter considers the principles of microwave solid-state amplifiers for the devices of Chapters 3, 4, and 5. Consideration is first given to the different classes of amplifiers and their desired class characteristics. Some devices serve better than others due to their low noise contribution during signal amplification. Some serve better as power amplifiers and can thus be designated as output amplifiers. When possible, a separation will be made between device types to permit specification of their optimum use.

6–2 THE GENERAL AMPLIFIER

Microwave amplifiers can be separated into three general classes, as follows:

(1) Low-noise amplifiers

(2) Linear amplifiers

(3) Output amplifiers

For the first two classes, certain amplifier properties are important. These are listed in Table 6-1.

Noise Factor
Gain
Bandwidth
Dynamic Range
Intermodulation Products

Table 6-1: Low Noise and Linear Amplifier Properties

For the third class, that of the output, or power, amplifier, other properties become important. Table 6-2 lists these properties.

Output Power
Gain
Bandwith
Power Generation Efficiency
Harmonic Content

Table 6-2: Output (Power) Amplifier Properties

From the tables, we see that bandwidth is important to all three classes of microwave amplifiers, as is the gain also. Noise factor (see below) is not of value to the output amplifier, as it must generally handle high signal levels. Efficiency is quite important to the output amplifier, as the drain on the power supply is usually heavy. Efficiency is not of prime interest in the low-noise amplifier, as the power supply drain is low and the signal power output is usually small compared to the input power needed.

Output amplifiers usually are low gain, but the output power level is high. The power generation efficiency η can be defined as follows:

$$\eta = \frac{P_{RF_{out}} - P_{RF_{in}}}{P_{dc} + P_{RF_{in}}},$$

where

$$P_{RF_{out}} = \text{rf output power}$$

$$P_{RF_{in}} = \text{input rf (drive) power}$$

$$P_{dc} = \text{dc input power.}$$

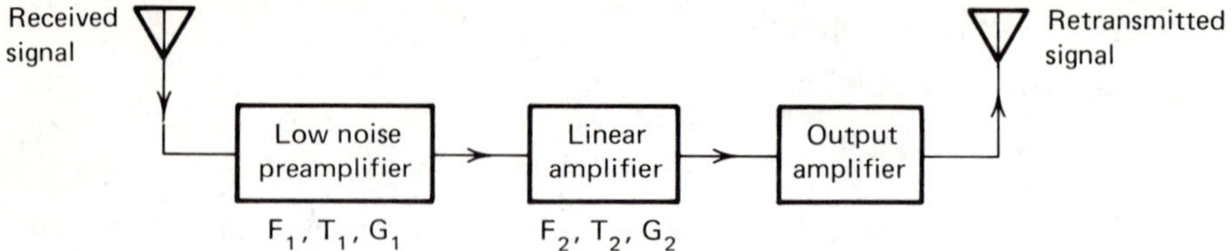

Figure 6-1: Communication Repeater Chain

A communication repeater could use all three types of amplifier classes, as in Fig. 6-1.

The repeater receives a weak signal from a distant station and amplifies it with a low-noise figure preamplifier. This is then amplified by a linear amplifier to provide enough signal to drive the output amplifier stage. The output amplifier then drives an antenna to retransmit the original signal in a relatively undistorted form without much additional noise added to the received signal.

The noise factor F is significant in the low noise amplifier, as it determines the smallest signal that can be detected. It can be defined as

$$F = \frac{\text{Total noise output}}{G \times \text{noise input}} = \frac{T_{eq}}{T_g},$$

where

$$G = \text{gain of amplifier}$$

$$T_{eq} = \text{equivalent temperature of total}$$
$$\text{noise power output } P_v = G\, k\, T_{eq}\, B$$

$$T_g = \text{equivalent temperature of}$$
$$\text{source generator.}$$

The noise output temperature can be approximated as

$$T_{eq} = T_a + T_g.$$

This means that a noise temperature T_a due to the amplifier itself is added to the noise temperature of the source generator T_g to determine the total noise temperature of the output signal. The noise power output is then given by

$$P_{noise} = k\, T_{eq} B \text{ watts,}$$

where

$$k = \text{Boltzmann's constant}$$
$$B = \text{Bandwidth (in Hz)}$$
$$T_{eq} = \text{output temperature in } {}^{0}\text{K}.$$

Referring to Fig. 6-1, we have the noise factor for the first two stages:

$$F_{12} = F_1 + \frac{F_2 - 1}{G_1},$$

where

$$F_{12} = \text{two-stage noise factor}$$
$$F_1 = \text{first-stage noise factor}$$
$$F_2 = \text{second-stage noise factor}$$
$$G_1 = \text{Gain of first stage}.$$

Clearly, with a reasonable gain G_1, the overall two-stage noise factor F_{12} will be determined essentially by the first-stage noise factor F_1.

6–2.1 Four-Terminal Amplifiers

The general four-terminal amplifier is given in Fig. 6-2. (With common input and output terminals, this would become a three-terminal device.) The amplifier has a power gain G, but also adds a noise P_n. The output load impedance (Z_1) has a reflection coefficient ρ_1 and a noise temperature T_1.

The total noise power output of the amplifier is then given by

$$P_N = GP_n (R_g, T_g) + P_n + |\rho_1|^2 P_n (R_1, T_1),$$

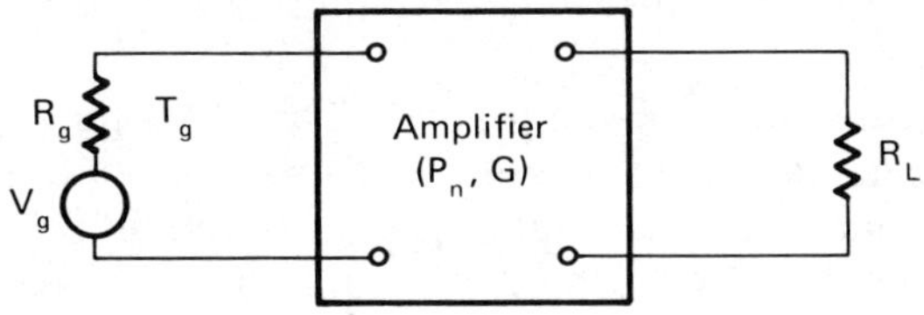

Figure 6-2: Four-Terminal Amplifier

The analysis usually performed using the circuit of Fig. 6-2 assumes a perfect amplifier with equivalent noise sources placed in the input circuit. In that case the amplifier noise can be written as

$$P_n(amp) = G \, k \, T_a \, B$$

A noise power of P_n/G placed in the input of a perfect, noiseless amplifier would give the same output noise power. Similarly, the noise from the input generator $P_n \, (R_g, T_g)$ is given by

$$P_n \, (R_g, T_g) = k \, T_g \, B.$$

The total noise output would then be given by $P_n = Gk \, T_{eq} \, B$. Using this philosophy, the equivalent temperature at the input of the perfect amplifier would be given by

$$T_{eq} = T_g + T_a + \frac{|\rho_1|^2 T_1}{G}.$$

The equivalent noise temperature at the input of the amplifier of Fig. 6-2 is then due to the temperature of the input generator T_g, the amplifier temperature T_a, and the load temperature T_1 reduced by the gain G.

Using the above concepts for a four-terminal amplifier, it is possible to predict the minimum detectable signal for any amplifier, given its source and load impedances, its gain G, the noise power added by the amplifier, and the degree of load mismatch (via reflection coefficient ρ_1). Clearly a perfect match ($\rho_1 = 0$) improves the low-noise performance by lowering T_{eq}.

6–2.2 Two-Terminal Amplifiers

The two-terminal devices of Chapters 4 and 5, when used as two-terminal negative resistance devices, can be used as amplifier devices. The devices can amplify if a microwave circulator is used to separate the input from the output circuits. Figure 6-3 illustrates two-terminal amplifier principles.

The circulator operates as follows. The input power P_{in} flows into port a and circulates around to port b. It flows out of b and, if reflected back, will flow into port b and out of port c. In the reflection process, the signal is amplified by $-R$ (at the diode), and the input and output signals are separated by the circulator action. For this reason, the device is also called a reflection amplifier.

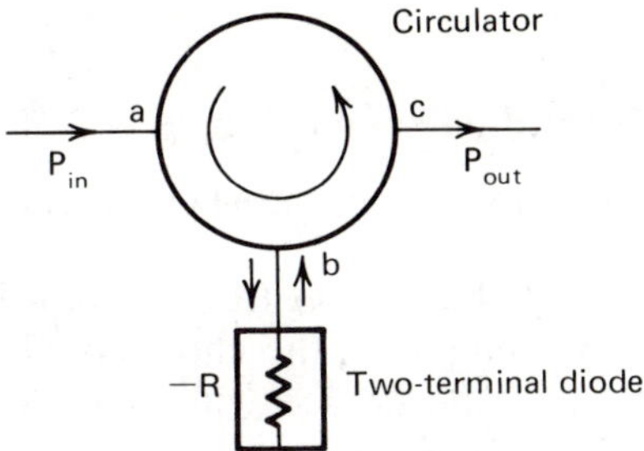

Figure 6-3: Two-Terminal Reflection-Type Amplifier

The Thevenin (constant voltage generator) equivalent circuit for the reflection amplifier is given in Fig. 6-4.

The gain for this device is given by the reflection coefficient at the plane of the device impedance Z_d at terminals a-a'. The reflection coefficient Γ is given by

$$\Gamma = \frac{Z_g}{Z_g^*} \cdot \frac{Z_d - Z_g^*}{Z_d + Z_g},$$

where

Z_g^* is the complex conjugate of the generator impedance.

The power gain is given by the reflection coefficient squared, and thus the gain G is given by

$$G = |\Gamma|^2 = \left| \frac{Z_d - Z_g^*}{Z_d + Z_g} \right|^2.$$

With proper impedance matching, the susceptance of the diode will cancel the susceptance of the load (generator), and the gain will be given by

$$G = \frac{(G_l + G_d)^2}{(G_l - G_d)^2},$$

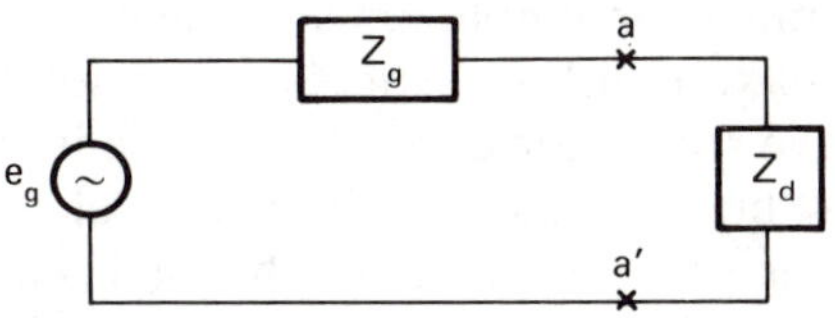

Figure 6-4: Thevenin Equivalent to Reflection Amplifier

where

$$-G_d = \text{diode conductance}$$

$$g_l = \text{load (generator) conductance.}$$

It would appear that the gain G (for a given diode) could become as large as desired, depending upon the choice of G_l. When $G_l = G_d$, the gain becomes infinite, and the device will oscillate. For stable amplification, the condition $G_l > G_d$ is imposed.

The analysis of the reflection amplifier is simplified if the Norton equivalent circuit is used. Figure 6-5 gives the Norton equivalent. The diode admittance is given by $Y_d = -G_d + jB_d$ and the load admittance by $Y_l = G_l + jB_l$. From this it is clear that susceptances (or parallel reactances) are cancelled at a-a' when $B_l = -B_d$. This condition is essential in order to have maximum gain.

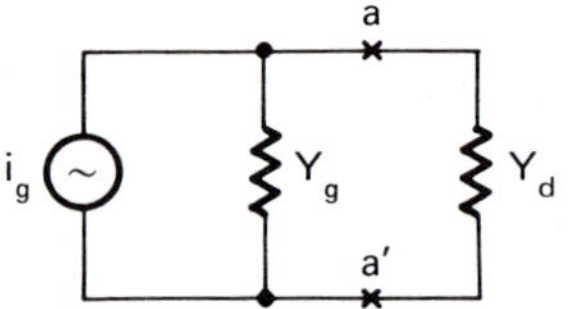

Figure 6-5: Norton Equivalent to Reflection Amplifier

Design of the microwave reflection amplifier requires careful impedance matching between the diode and the waveguide on the generator side. The diode's admittance is transformed by the waveguide before it reaches terminals a-a', and the load on the diode is also a transformed version of the generator admittance.

6–3 BIPOLAR AND UNIPOLAR MICROWAVE AMPLIFIERS

Microwave transistor amplifiers in general require matching networks to achieve maximum gain. This is suggested by Fig. 6-6.

The microwave transistor current equivalent circuits of Fig. 3-2 of Chapter 3 can now be inserted into the active element box of Fig. 6-6. The impedance of the generator (Z_s) and the load (Z_l) can now be adjusted by the networks N_1 and N_2 to achieve maximum unilateral (one way) gain. The conditions for achieving this will not be discussed here, but it can be simply said that a technique exists that will permit the design of an op-

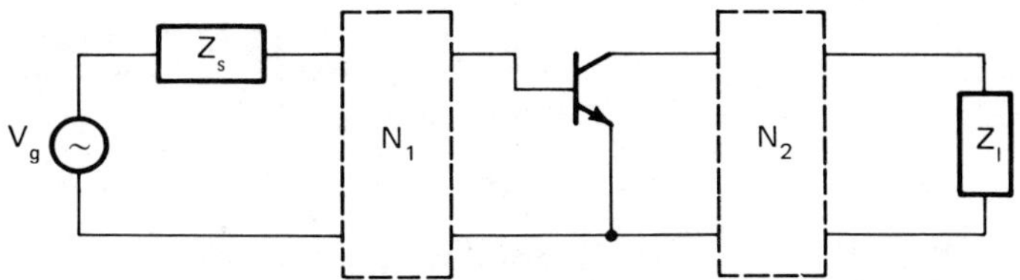

Figure 6-6: Microwave Transistor Amplifier

timized amplifier when the transistor's S-parameters are measured by a network analyzer system. The forward power gain is maximized when the input and output networks are conjugately matched. The condition of having a negative resistance in either the input or output ports (or both) will permit oscillation, and thus this condition is in general avoided.

The equivalent circuit of the unipolar MESFET of Chapter 3 may also be inserted into the active element box of Fig. 6-6 to permit design of an optimized MESFET amplifier. Again, the matching networks N_1 and N_2 are designed to conjugately match the input and output networks over the frequency ranges of interest.

The detailed MESFET equivalent circuit is given in Fig. 6-7 for reference. (See Chapter 3 for MESFET principles of operation.)

Clearly, a detailed analysis is needed in order to optimize the unipolar microwave amplifier design.

6–4 TED (GUNN) AMPLIFIERS

6–4.1 Stabilizing Considerations

In Chapter 4 the transferred electron device (TED) was discussed and its operation described. The TED is basically a two-terminal oscillator device. The negative differential mobility of the device can permit its use as an amplifier, assuming that small, instantaneous disturbances are not allowed to form in the device. Such space charge disturbances in the device would grow due to the negative differential mobility, making amplification impossible.

The following techniques stabilize the Gunn device to prevent self-oscillation:

n_0L *Stablilization.* It can be shown that if the product of the impurity density n_0 and the length L is less than $10^{12}/cm^2$, then the diode is stable and will not go into self-oscillation. This principle works only if the proper load impedance is connected to the device.

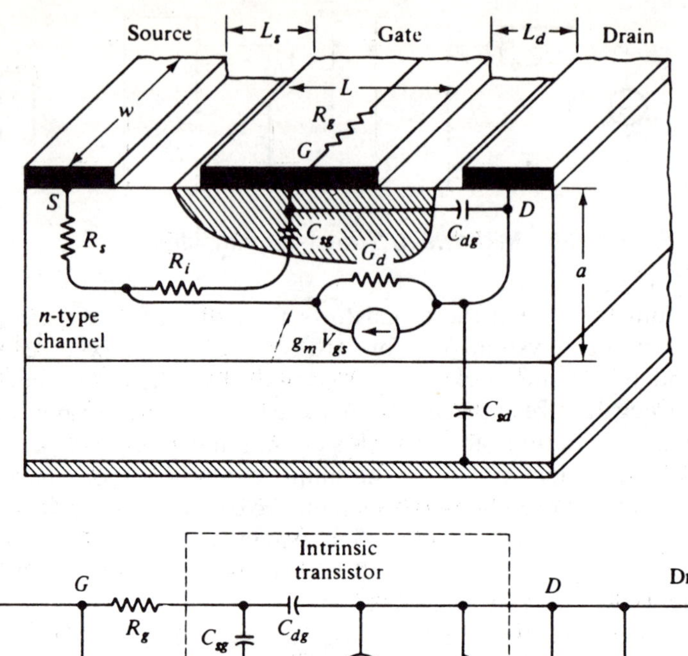

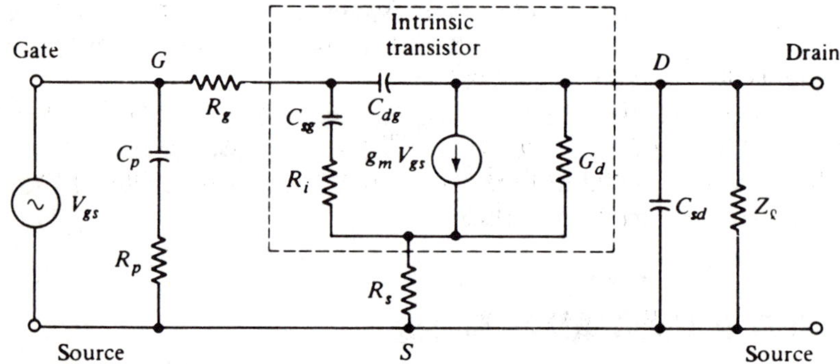

Figure 6-7: MESFET Equivalent Circuit *

Circuit Stabilization. A diode with an overcritical $n_0 L$ product can be stabilized by the addition of enough positive resistance to cancel out the negative resistance. The technique can only apply to cases where the $n_0 L$ product slightly exceeds the critical value. The reason for this is that any resistance added also provides additional susceptance to the microwave circuit, adding instability.

Diffusion Stabilization. In cases where the $n_0 L$ product exceeds the critical value, small space charge disturbances can grow. However, they may form a new field distribution that is stable. Certain n_0 values have been found that provide stability. However, with large signals the device may go into oscillation, and Gunn oscillations will form.

Notch Stabilization. An amplifier can be made with an overcritical $n_0 L$ value but with a narrow region of reduced doping near the cathode. The notch causes a high electric field, but this high field remains constant throughout a major portion of the diode. The high field leads to an effective increase in the critical n_0L product, producing stabilization.

6–4.2 Noise Level

TEDs are not promising for use as low-noise amplifiers, since hot electrons are present, and it is these that are the effective producers of thermal noise power. The total noise power generated by these amplifiers has been calculated by Shockley and his co-workers. GaAs amplifiers with noise factors under 10 dB have been developed. InP amplifiers with noise factors of about 11 dB have been tested at 8.5 GHz.

6–4.3 Intermodulation Characteristics

At large rf signal drive levels, several nonlinear effects occur in the Gunn device. One of these is intermodulation distortion. This type of distortion can be measured by injecting two single frequencies into the device, separated by a few MHz, and observing the output on a spectrum analyzer. A series of products will appear on the spectrum analyzer.

Figure 6-8 illustrates the kind of display that can be expected; f_3^+ and f_3^- are the third-order upper and lower intermodulation distortion products. A perfect amplifier would have no measurable intermodulation products. With increased signal drive and approach to saturation of the device, these higher-order distortion products will increase in magnitude. A typical amplifier will exhibit the kind of intermodulation characteristics shown in Fig. 6-9.

In the figure, as the input power is increased, a level will be reached (point A) where the third-order distortion products become appreciable.

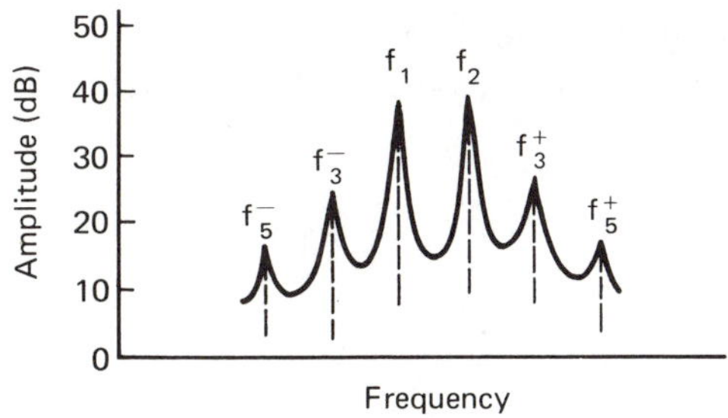

Figure 6-8: Form of Intermodulation Spectrum Display

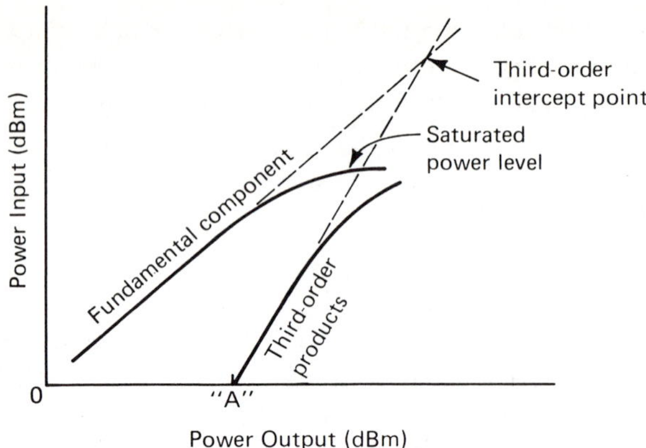

Figure 6-9: Stable Amplifier Intermodulation Characteristics

This is certainly true in the beginnings of the nonlinear portion of the input-output fundamental component curve. The third-order intercept point shown is about 6-10 dB above the saturated power output level for Gunn devices.

The output power of these devices is approximated by the formula

$$P = E_1^2 \, |\mu| \, e \, A \cdot n_0 L,$$

where

$$E_1 = \text{average ac field swing}$$

$$A = \text{diode cross-sectional area}$$

$$n_0 = \text{doping density}$$

$$L = \text{diode length}$$

$$e = \text{electron charge}$$

$$\mu = \text{mobility.}$$

The output power is thus proportional to both the product n_0L and the cross-sectional area of the diode.

The product of the power output of the diode and its impedance is proportional to $1/f^2$, or

$$P \cdot Z \, \alpha \, E_1^2 L^2 = E_1 \, V_0^2 \, \frac{1}{f^2},$$

where

$$v_0 = \text{electron velocity.}$$

For this reason, the diode length must be shortened for higher-frequency operation.

6–5 AVALANCHE INJECTION AMPLIFIERS, IMPATT AND TRAPATT

6–5.1 The TRAPATT Amplifier

The IMPATT and TRAPATT diodes of Chapter 5 can be used as microwave amplifiers. The TRAPATT diode is a large-signal device which oscillates at the IMPATT frequency. It can oscillate at the IMPATT frequency while oscillating or amplifying at a lower frequency. The TRA-PATT amplifier exhibits a very high noise level (60 dB), probably because of the time jitter of the avalanche "ignition" process.

The TRAPATT as an oscillator has very high efficiency in the pulse mode (60%). However, the power density in the device ranges from 10^5 to 10^6 w/cm^2, preventing cw operation for reasonable junction temperatures. Above 10 GHz, no useful TRAPATT power can be obtained. An IMPATT oscillation must be started before TRAPATT oscillation can be initiated. This latter fact is cited as the main reason for limiting the TRA-PATT mode to under 10 GHz. Because of the noise level of the TRAPATT mode and its declining interest in the microwave solid-state community, it will not be discussed further.

6–5.2 The IMPATT Amplifier

6–5.2.1 IMPATT principles

As discussed in Chapter 5, the IMPATT diode was proposed by Read in 1958. With the proper choice of drift length L, the terminal current can be made to be 180° out of phase, and the diode will produce a negative resistance at half the transit-time frequency f_T, or at

$$f = \frac{f_T}{2} = \frac{v_s}{2L}$$

and

$$v_s = \text{saturation velocity}$$
$$L = \text{diode length.}$$

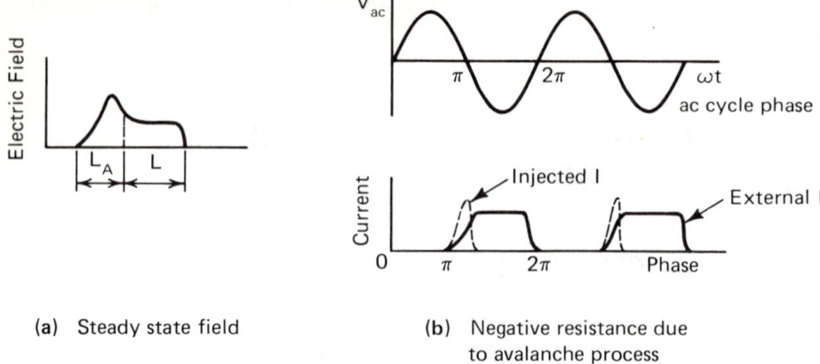

Figure 6-10: Field and Current in a Read-Type Diode

The static field in an IMPATT diode is given in Fig. 6-10a. During the ac cycle between $\pi/_2$ and π radians, the high-field region in the diode generates carriers. Current is injected into the device during the most rapid decrease of the applied voltage, giving a 90° phase lag with the applied voltage. The net result is an external current pulse that is 180° out of phase with the applied voltage, as indicated in Fig. 6-10b.

6–5.2.2 Amplifier considerations

The IMPATT as an amplifier has a limited range of ac voltage swing, due to the necessity of keeping the carrier energy levels below breakdown. This limits the power and efficiency of the amplifier in large-signal operation.

The amplifier must remain stable, and thus the impedance of the diode (which will have a negative resistance and a capacitive reactance) combined with the load impedance must lead to a positive overall impedance over the frequency range of operation. The real part of the diode impedance (i.e., its resistance) must remain negative over the frequency range in order to achieve gain.

The IMPATT diode ceases to have a negative resistance above the resonant frequency F_a and thus cannot amplify above this frequency. This limits the diode to a one-octave frequency band.

Figure 6-11 shows the equivalent circuit of the diode in the avalanche region. The parallel resonant frequency of this region is given by

$$f_a = \frac{1}{2\pi \sqrt{L_A C_A}}.$$

This frequency is proportional to the square root of the dc current density J_0.

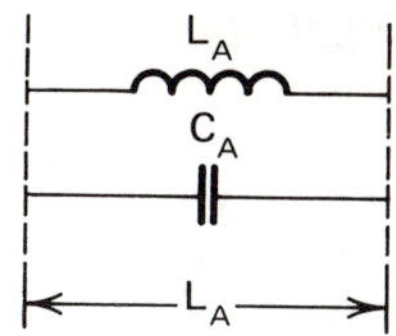

Figure 6-11: IMPATT Diode Impedance

6–5.2.3 Noise, saturation power, and efficiency of IMPATT amplifiers

The noise level of an IMPATT amplifier is very high, but so are the power levels and the efficiency. High gain and adequate bandwidths have also been obtained, making the amplifiers adequate for many microwave system needs. However, careful bias circuit design is necessary to avoid instabilities. Parametric or triggered oscillation can occur, along with large excess noise and intermodulation effects.

Because of the high noise levels, IMPATT amplifiers are used exclusively as power amplifiers. Figure 6-12 is an example of an IMPATT amplifier circuit. The IMPATT diode is coupled via coaxial line segments to a microwave cavity. The coaxial termination used for coupling in the dc bias current is broadband, which tends to provide proper loading to avoid low-frequency instabilities (i.e., in the MHz range).

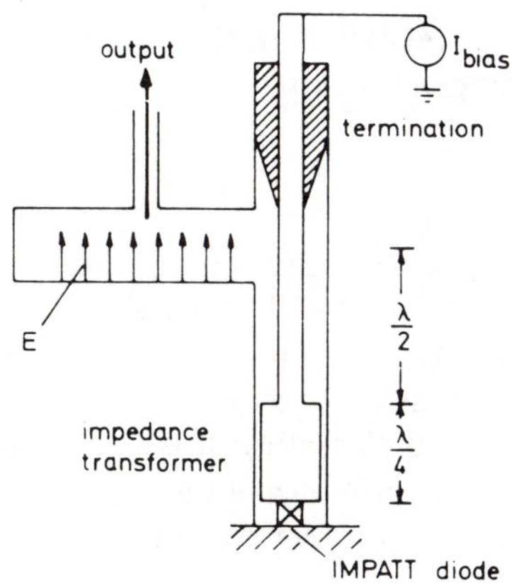

Figure 6-12: A Practical Coaxial IMPATT Amplifier Circuit *

6–6 THE BARITT AMPLIFIER

The BARITT device of Chapter 5 can also be used as a microwave amplifier. As seen in Fig. 5-10 of that chapter, the BARITT is a back-to-back combination of two p-n junctions, one being forward biased and the other reverse biased. The device is biased to near punch-through conditions, and an input signal drives it periodically into that region.

The frequency of the BARITT is given by

$$f = 0.75 \, V_s/L,$$

where

$$V_s = \text{saturation velocity of carriers}$$

$$L = \text{diode length.}$$

Since the phase delay between current and voltage in the device is between 90° and 180°, the real part of the diode impedance is negative, but a strong capacitive reactance also appears. This latter quantity reduces the bandwidth of the device.

The small-signal impedance of the device in combination with the load impedance must remain in the stable limits of operation for the frequency range where the real part of the device is negative.

6–7 SOLID-STATE AMPLIFIER
DESIGN CONSIDERATIONS

An interesting application illustrating the principles outlined in this chapter is a hybrid TED-IMPATT design. The TED amplifier has low noise properties and high gain, and thus the TED is suitable for the input stage. The TED also has a wider bandwidth than the IMPATT amplifier. The IMPATT can be used as both a driver and an output amplifier and can have a larger overall bandwidth at the sacrifice of its single-stage gain. Figure 6-13 is the diagram of a five-stage hybrid amplifier using TED and IMPATT stages.

It is interesting that the efficiency of the hybrid amplifier improves with higher input power. (See Fig. 6-14.) The important point is that

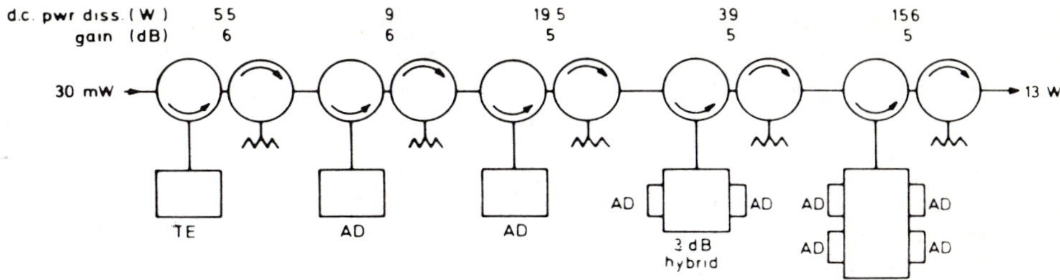

Figure 6-13: Five-Stage Hybrid Amplifier (TE = Transferred Electron, AD = Avalanche Diode)*

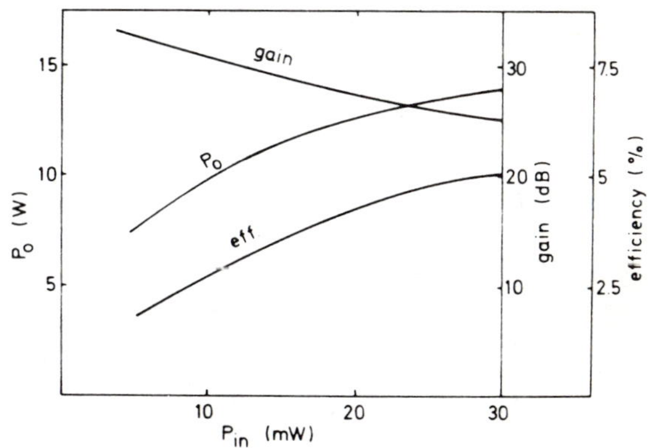

Figure 6-14: Characteristices of Hybrid Amplifiers*

several types of solid-state amplifiers can be coupled together so that a more optimum overall amplifier system can be achieved.

A comparison can be made between TED, BARITT, and IMPATT amplifiers. Table 6-3 compares these three devices when they are used as amplifiers.

PROPERTY	**COMPARATIVE RANKING**
1. Impedance Level	BARITT < IMPATT < TED
2. Power	BARITT < TED < IMPATT
3. Noise	BARITT ≤ TED < IMPATT
4. Gain-bandwidth product	BARITT < IMPATT < TED
5. Efficiency	BARITT < TED < IMPATT

Table 6-3: Comparison of BARITT, TED, and IMPATT Amplifiers

---

*Microwave Devices, M.J. Howes and D.V. Morgan, © 1976, John Wiley & Sons, Ltd. Reprinted by permission of John Wiley & Sons, Ltd.

Hybrid combinations of microwave solid-state amplifiers other than that of Fig. 6-13 are possible. A very low noise input stage might make the BARITT a good choice, but its low gain bandwidth product would serve as a limitation. A lower bandwidth would permit a higher sensitivity of the BARRITT as a preamplifier and would not demand that an IMPATT output stage be operated at low gain. Table 6-3 can, then, be used as a guide to other hybrid amplifier chain designs, using the principles outlined in this chapter.

7 Plasma Wave Electronics

7–1 INTRODUCTION

In the chapters considered so far, solid-state devices have been prominent. In Chapter 5 the TRAPATT oscillator was considered, and that device contained what was called a trapped plasma. (Recall from Chapter 5 that TRAPATT is the acronym for *trapped plasma triggered transit*.) This chapter considers possible devices involving waves in plasma, a phenomenon which has existed in nature for a long time but has only in recent decades been explored.

7–2 PLASMA IN NATURE

Plasma has been called a fourth state of matter. Consider the following process. A metal is melted, and it naturally becomes a liquid, having changed from its original solid state. The liquid (molten) metal is heated again, and it is changed into a metallic vapor or gas. This is its third state. Consider now adding more heat energy so that the electrons of the gas atoms (or molecules) are stripped away and the vapor is now broken up into gas atoms (or molecules) minus their electrons (positive ions) plus the free electrons (negative ions). Figure 7-1 illustrates these four stages for the case of the metal copper. The net charge in this mixture of ions will be near zero. The positive ions will have nearly the same density as the negative ions, giving the net neutral charge. This neutral gas of charged particles is called a *plasma*. For distances in the plasma greater than a fundamental length called the *Debye length*, the plasma

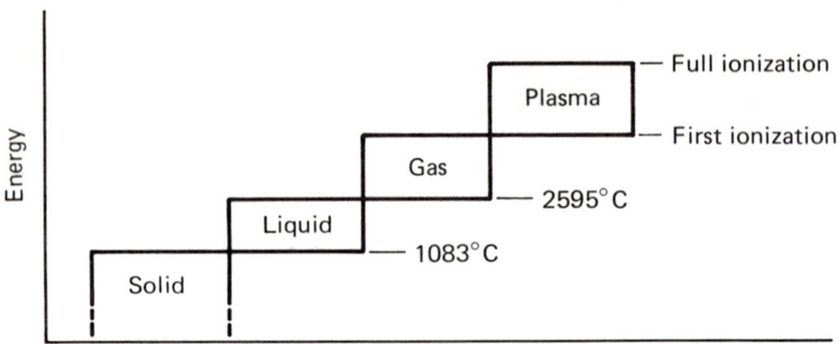

Figure 7-1: Example of the Four States of Copper

will appear to be a homogeneous, uniform mixture of the two ion types.

Plasmas occur in nature. The matter of the universe is made up mostly of plasmas. The stars are examples of objects whose temperatures are high enough to have their elements in a plasma state. Plasmas also exist in the ionosphere, a layer of ionized gas located between about 100 and 350 km above the earth's surface. This ionospheric plasma causes the bending (refraction) of hf radio waves so that they can travel over long distances. The plasma of the ionosphere makes it possible to use short-wave radio. Plasmas also exist in the northern latitudes and are associated with the Aurora Borealis (The Northern Lights). These intense plasmas also contain large currents that travel roughly along lines of equal latitude above what is termed the aurora oval.

Thus, the plasma state is common in nature, a fact that is widely known. It can occur in a solid-state material, as evidenced by TRAPATT experience, as well as in metals or gases. Over 99% of the universe is in the plasma state.

7-3 BASIC PLASMA CONCEPTS

Consider now the significant plasma parameters, such as density, temperature, and the like. The density of the plasma is usually given in particles per cubic centimeter. It is thus a number density. The term used commonly is n_e, for number of electrons per cubic centimeter (cm^3). Another useful quantity is what is called a plasma frequency. This is the natural oscillation frequency of the plasma. In MKS units the plasma frequency is written as

$$\omega_p^2 = \frac{n_e q^2}{m\,\epsilon_0} = (2\pi f_p)^2,$$

or

$$f_p^2 = \frac{n_e q^2}{4\pi^2\,m\,\epsilon_0},$$

where

$$n_e = \text{electron (number) density}$$
$$m = \text{electron mass}$$
$$q = \text{electronic charge}$$
$$\epsilon_0 = \text{permittivity of free space.}$$

In approximate values,

$$f_p \simeq 9000\ \sqrt{n_e}$$
$$n_e = \text{electrons/cm}^3,$$
$$f_p \text{ in Hz.}$$

Clearly, the higher plasma frequency values always require high electron densities.

The plasma behaves like a dielectric in some ways. Its dielectric constant can be approximated by

$$\epsilon_r = 1 - \frac{\omega_p^2}{\omega^2},$$

where

$$\epsilon_r = \text{relative dielectric constant}$$
$$\omega/2\pi = \text{incident radio frequency on plasma}$$
$$\omega_p = \text{local plasma radian frequency}$$
$$= 2\pi\,f_p.$$

An important parameter is the ratio of the speed of light to the phase velocity of an EM wave in the plasma. This is the index of refraction η. For a low-loss plasma, η is given by

$$\eta = \sqrt{\epsilon_r} = \left(1 - \frac{\omega_p^2}{\omega^2}\right)^{1/2} = \frac{c}{V_{ph}},$$

where

$$V_{ph} = \text{phase velocity in plasma}$$

$$c = \text{speed of light.}$$

In the above formulas the effects of collisions between electrons and neutrals (or ions) are neglected.

The phase velocity is usually expressed as

$$V_{ph} = \frac{1}{\sqrt{\mu\epsilon}} = \frac{1}{\sqrt{\mu_0\epsilon_0}\sqrt{\epsilon_r}} = \frac{c}{\sqrt{\epsilon_r}}.$$

In the plasma,

$$V_{ph} = \frac{c}{\sqrt{\epsilon_r}} = \frac{c}{\left(1 - \frac{\omega_p^2}{\omega^2}\right)^{1/2}}.$$

Three unusual things happen here that do not occur in ordinary material. First, a negative dielectric constant can occur (when $\omega < \omega_p$); second, the phase velocity in the plasma can exceed that of a light wave; and third, the index of refraction can become imaginary.

Since the phase velocity varies with the radio frequency ω, the plasma is termed *dispersive*. In a dispersive medium, waves of different frequencies have different phase velocities and thus different wavelengths. The general relationship between phase velocity and wavelength in any medium is

$$\lambda = \frac{V_{ph}}{f}.$$

Because the medium is dispersive in a plasma, there is no simple relationship between wavelength and frequency. Remove the plasma, and the EM wave travels at the speed of light and $\lambda = \frac{c}{f}$, where c is the speed of light. This formula is used to convert from frequency to wavelength. A frequency of 30 GHz, then, corresponds to a wavelength of 10 mm. The above statement is not true for a plasma wave. The more precise statement for a low loss (or zero loss) plasma is

$$\lambda = \frac{c}{f} \cdot \frac{1}{(1 - f_p^2/f^2)^{1/2}} \cdot$$

At an EM frequency f equal to the plasma frequency f_p, the denominator goes to zero, and the wavelength λ goes to infinity. At a frequency f less than f_p, the wavelength is imaginary. (Recall that the square root of a negative number is imaginary. The number under the root sign for $f < f_p$ is negative, giving the condition of a negative square root in the denominator.)

Figure 7-2 illustrates the case of an electromagnetic wave of frequency f striking a slab of uniform plasma with dielectric constant ϵ_r. The EM wave can interact in three different ways, as indicated in Table 7-1.

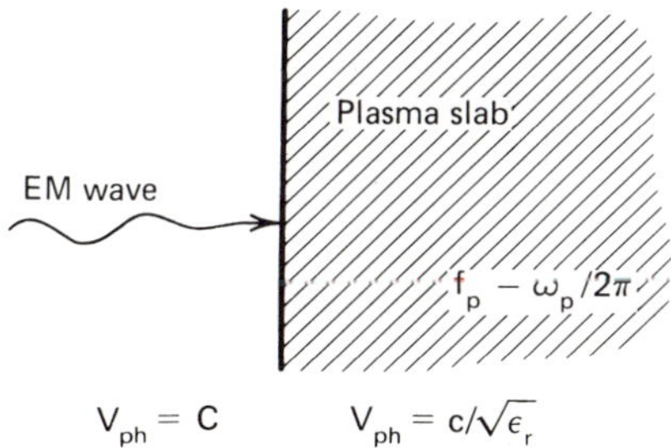

Figure 7-2: EM Wave Incident on Plasma Slab

If the incident wave is at a frequency f greater than f_p, then it penetrates the plasma slab, but must operate in a dispersive medium. For the case of f equal to f_p, the wave is reflected in a manner similar to the way it would be reflected by a metallic plate. For a frequency f less than f_p, the EM wave is bent (refracted) when the EM wave comes in at an angle to the plasma slab face. This refraction is indicated in Fig. 7-3, for the case of high frequency (HF) waves in the ionospheric plasma. The refractive effect is what makes it possible to transmit and receive HF radio waves at great distances.

Frequency Regime	$f < f_p$	$f = f_p$	$f > f_p$
Characteristics	EM wave reflected, essentially no penetration	Critical case for EM wave. Reflects like a metal	EM wave propagates, but medium is **dispersive.**

Table 7-1: EM Wave Interaction with Plasma Slab

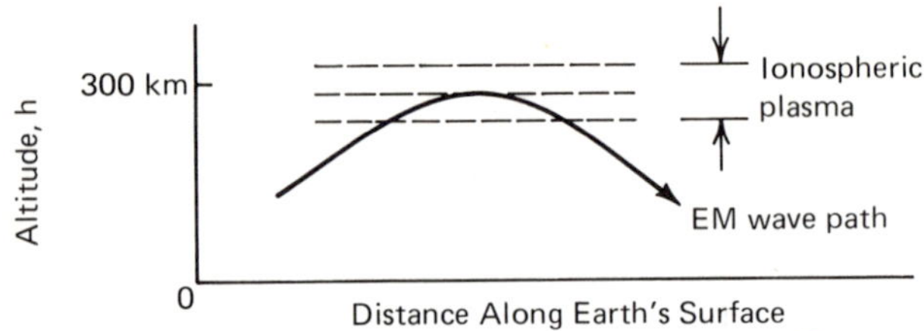

Figure 7-3: Ionospheric Plasma Refraction of an EM Wave

7–4 WAVES IN THE PLASMA

The previous section described how an EM wave interacts with a uniform plasma. What actually happens is the electric field of the incoming EM wave causes the electrons in the plasma to oscillate. The natural oscillation frequency of the plasma is $\dfrac{\omega_p}{2\pi}$, so that when the EM wave's frequency is near to $\dfrac{\omega_p}{2\pi}$ there is a natural coupling between the wave and the plasma, with the greatest interaction at the resonant condition $\omega = \omega_p$. The plasma can be viewed as a form of dielectric material with a dielectric constant ϵ_r. This picture of a plasma is very incomplete, as it does not allow for plasma waves, i.e., waves that are not EM in nature, or waves that can either grow or decay as they propagate in the plasma.

The plasma itself is composed of electrons that are dispersed over a range of velocities. The velocities of the electrons are distributed statistically over a given distribution, the most common being the Maxwellian distribution. There exists within the plasma a number of natural or "normal" modes that can permit the growth of longitudinal electric waves in the plasma. The energy for these modes will be supplied by the velocities of the electrons themselves. A certain range of electron velocities will stimulate these modes, and the waves will grow out of the thermal levels in the plasma. It seems that these wave modes normally occur in the plasma at room temperatures, being driven by the energy (electron velocities) in the plasma itself.

It can be shown that, for a collisionless plasma, these waves grow at the Langmuir frequency, which is the plasma frequency ω_p. The waves grow when electron velocities exist in a direction $\hat{k}$ that coincides with the direction of the wave mode in the plasma. This particular electron velocity must be at the phase velocity of the wave

$$V_{ph} = \frac{\omega}{k},$$

where

$$\omega = \text{wave frequency}$$

$$= 2\pi f$$

$$k = \text{wave number}.$$

Electrons whose velocities exceed V_{ph} will give up energy to the wave, and electrons whose velocities are less than V_{ph} will obtain energy from the wave. Figure 7-4 illustrates what can happen for two types of velocity distribution functions f(v). Figure 7-4a is the normal, or Maxwellian, case. With electrons distributed as shown, and with the mode phase velocity v_{ph}, there are more electrons that receive energy from the mode than give energy to the mode in the plasma. (This effect will be discussed further in Chapter 9. Figure 9-6, using a potential well analogy,

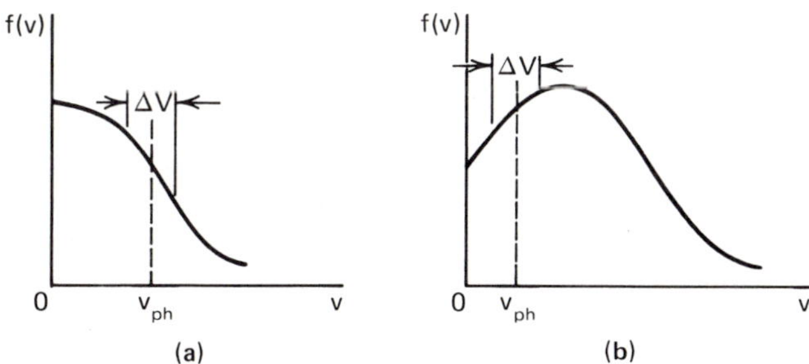

Figure 7-4: Velocity Distribution Functions

will show how an electron of velocity v_0 could feed a growing wave in the electron stream of a travelling wave tube. The same mechanism is used here, but with a normal plasma mode in place of the bunched electron stream.) Figure 7-4b gives a velocity distribution function f(v) that has a region that will permit a normal mode of velocity v_{ph} to grow. The positive slope of f(v) in that region permits wave growth, since there are simply more electrons with velocity greater than v_{ph} in the vicinity of v_{ph}. This results in feeding energy from the electron distribution function into the normal mode with the phase velocity v_{ph}. Now if this also occurs along the direction of the mode k, then these modes will grow out of the thermal background in the plasma. They will grow until the distribution f(v) is depleted to an equilibrium value f'(v), as illustrated in Fig. 7-5.

When the electrons lose energy to the wave mode, they lose velocity and fall back into a lower position in the f(v) distribution. The distribution

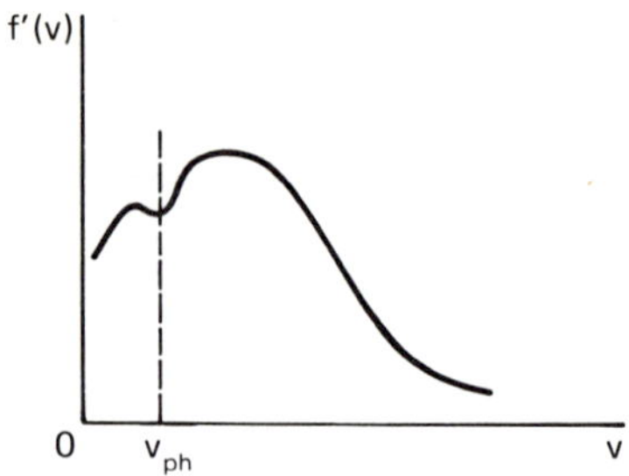

Figure 7-5: Depletion of f(v) Due to Wave Absorption Effects

functions of Fig. 7-4b and Fig. 7-5 can be approximated by displaced Maxwellian distributions. These can be formed by injecting a current flow into the plasma, so that the plasma will have an electron stream slipping through the positive ion background. This is what is called the *two-stream effect* in a plasma. This can also be formed by injecting two electron streams into the plasma in opposite directions.

Consider now a general distribution of electrons in the plasma that is non-Maxwellian and has a tail in the distribution. Figure 7-6 shows a general case.

In this case, waves with lower phase velocities in Region A could be stimulated by the electrons of that region, and waves with phase velocities in Region B could also be caused to grow. It should be emphasized here that this is a nonequilibrium case for f(v), one that must be maintained by external driving electron forces and streams. The curve can be described as a drifted Maxwellian distribution with a bump in its tail. It could permit driving two separate modes—the acoustic mode of Region A and the optical mode of Region B—in the plasma. Region A's mode would require v_{ph} near the velocity of acoustic waves, and Region B's mode would require phase velocities exceeding or nearly equal to the electron thermal velocity.

It is clear, then, that "warping" the velocity distribution function of a plasma can supply mechanisms for growth of the normal modes that

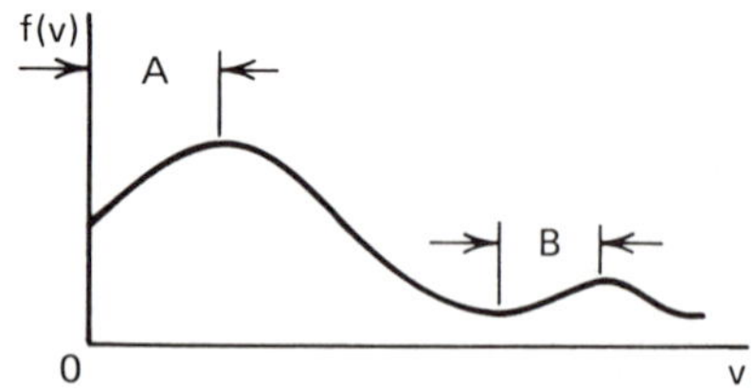

Figure 7-6: Dual-Region Distribution Function for Growing Waves

exist in the plasma. The wave growth would then be limited by saturation effects or the tendency to deplete the distribution function of local regions that feed the normal waves. Since the velocity distribution function f(v) must be "warped" in the direction where the normal mode occurs in the plasma, some special tailoring of f(v) is required. Mathematically, f(v) then become f(v, r̄), emphasizing that this function will be different for different directions in both position and velocity space. Strictly speaking, there must be a matching of the particle distribution function in position and velocity spaces with the normal mode in k-space in the plasma. In the section that follows, an example is given from nature in which this actually occurs on a daily basis, showing that it is a real phenomenon.

7–5 GROWING WAVES IN THE AURORA BOREALIS

Radar experimenters operating in the northern latitudes have directed their equipment toward the aurora borealis and noticed special reflections from the aurora. The radar reflections (or "radar returns") would increase in strength as the auroral electrojet current became stronger, showing correlations with the electrojet amplitude. This current system in the aurora has been estimated as building up to a million amperes (10^6 A) every day, flowing in a general east-to-west (longitudinal) direction at an altitude close to 100 km. This altitude is in the ionospheric E-region of the atmosphere, a region known for its sporadic radio reflection properties. Considerable radio and radar interference has been generated by the aurora, and strong reflections are known to occur due to these Northern Lights.

The present understanding of the mechanisms responsible for this reflection property of the auroral plasma is that electrostatic waves build up in this plasma, and these waves reflect the EM energy of the radio or radar system. At least part of the time, these reflections occur coincidently with the buildup of the auroral electrojet. As mentioned in the previous section, a current system in a plasma can distort the velocity distribution of the electrons so that plasma waves can be stimulated and grow. What is suspected is that a mechanism for spontaneous and stimulated emission, similar to that of the laser, can occur. (See Sect. 10.2 and Fig. 10-3.) The conditions for stimulated emission can be applied to the nonequilibrium plasma to determine what range of parameters would be required to permit a laser-type wave growth in the plasma. The distorted velocity distribution would then be equivalent to the population inversion in the laser medium.

Rocket probes launched into the auroral plasma tend to show build-ups of electron density fluctuations with the electrojet that would coincide with strong electrostatic wave growth, i.e., the growth of plasma waves.

7–6 EXPERIMENTAL EVIDENCE FOR THE GENERATION AND CONTROL OF LONGITUDINAL PLASMA WAVES

Can plasma waves be generated and controlled? Many experimenters have generated plasma waves and measured their existence. Few, however, have tried to exercise gradual, continuous stimulation and control of them. The author wrote* a dissertation on the control of these plasma waves, showing how they could be suppressed by modification of the electron velocity distribution function. Large simulation experiments were set up in which a scaled representation of the E-region of the ionosphere was developed in the laboratory to represent auroral type plasma background conditions, and an electrojet was introduced into the medium. When the electrojet current level exceeded the critical velocity, plasma wave buildup was observed in the EM wave scattering experiments. The experimental arrangement is shown in Fig. 7-7. The scaling used was based upon the electrical conductivity of the plasma. Very high electronic and ionic temperatures were required in the scaling, and high microwave scattering frequencies were needed in order to approximate the proper wave number interactions. The theory predicted that once the instability was generated, it could be suppressed by "thermalizing" the electrons.

Figure 7-8, from the said dissertation, shows two computed marginal-stability boundary curves. The normal arc discharge in helium operates at the value indicated on the graph of Fig. 7-8b, marked "A." (The two curves show the effects of ion-neutral collisions on the marginal-stability boundaries. The term $\omega_i \tau_i$ is the product of the ion plasma frequency and the mean time for collisions between ions and neutral particles in the plasma.) The value is within the region of growing waves on the figures, and thus the waves can build up spontaneously. They will also have the wave number spectrum width† indicated on the figures. With the introduction of more thermal energy to the electrons, their thermal velocity V_- will increase. The ratio of the drift velocity V_d to V_- will then decrease, the operating point will move to the left, and the operation will be outside of the marginal stability boundary. This means that electrostatic waves will no longer grow in the plasma and the plasma will become like

---

*J.T. Coleman, "The Transition to Instability in a Nonequilibrium Plasma", University of Microfilms, Ann Arbor, Mich., 1967.

J.T. Coleman, "Control of Ion-Acoustic Waves in a Gaseous Discharge", Journal of Applied Physics *38*(6), pp. 2655-2659 (1967).

†Spectrum width $= \dfrac{\Delta k}{k_e} = \dfrac{1}{k_E}(k_2 - k_1)$.

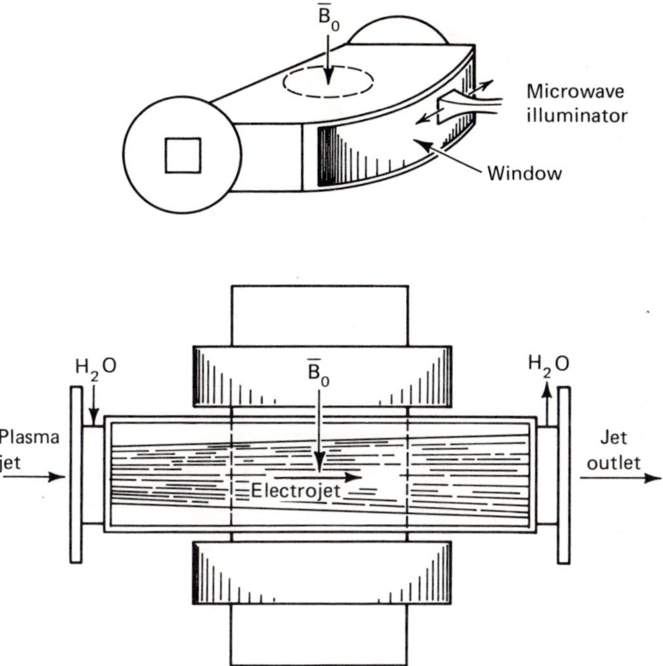

Figure 7-7: Auroral Simulation Experiment

the plasma slab model of Fig. 7-2. For his dissertation, the author generated the conditions of Fig. 7-8 in a hydrogen plasma, with an electron-to-ion temperature ratio of approximately 19. The arc discharge in hydrogen had the electron-drift-velocity-to-thermal-velocity ratio as indicated. EM wave forward scatter measurements (at 9 GHz) showed a spectrum of waves introduced by the turbulence of the wave vectors, and this extended roughly over the given k_1 to k_2 range. (See Fig. 7-9 for a diagram of the scattering experiment.) The plasma, with full drift current, was heated by an rf source to impart more thermal energy to the electrons. This increased the term V_- in the V_d/V_- ratio, lowering the ratio. This in turn moved the main operating point on Fig. 7-8 to the left, gradually reducing the spectral width $k_2 - k_1$ and dropping the turbulence spectrum down into noise. This corresponded to experimentally moving across the marginal stability boundary into the decaying wave region (the cross-hatched region of Fig. 7-8).

The experiment showed that electrostatic waves could be spontaneously stimulated in a plasma, grow over the region of wave numbers predicted, and be suppressed as required by the theory. The secret was the modification of the electron velocity distribution function to permit

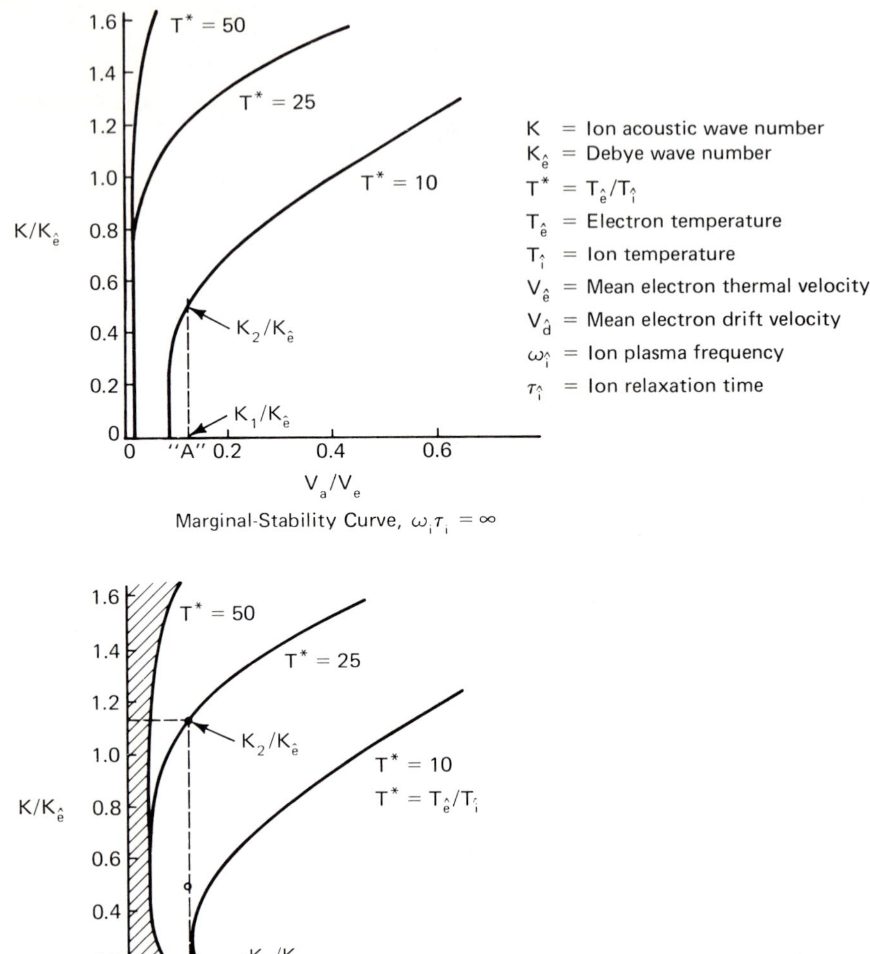

Marginal-Stability Curve, $\omega_i \tau_i = \infty$

Marginal-Stability Curve, $\omega_i \tau_i = 100$

Figure 7-8: Marginal-Stability Boundary Curves

control over the stimulation and growth of the normal-mode electrostatic waves in the plasma. In the normal Maxwellian plasma, these waves always exist. When permitted, they will spontaneously grow and change the plasma from a somewhat passive dielectric slab to a medium for growing waves.

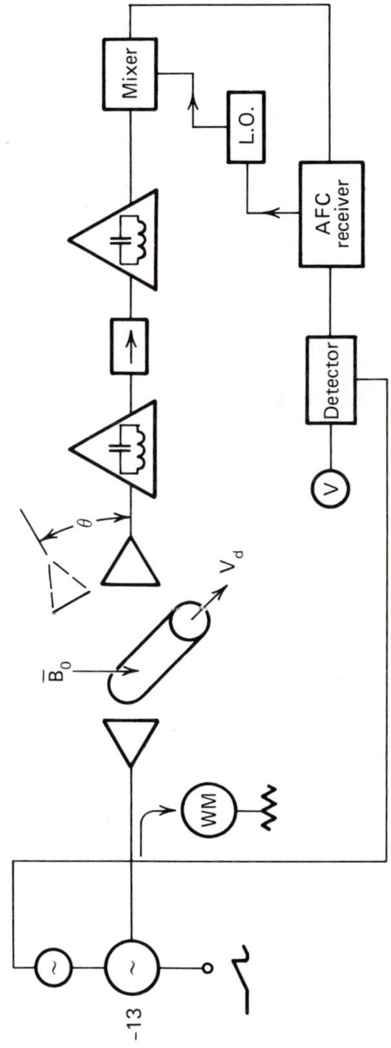

Figure 7-9: Plasma Experiment Configuration

7-7 POSSIBLE PLASMA WAVE DEVICES

Plasma waves have been shown to exist in a plasma medium and also have the ability to interact with EM waves. Under the proper conditions, they may be caused to grow or decay in certain directions in the plasma. When an EM wave of the right frequency strikes the plasma, it causes the electrons in the plasma to oscillate (vibrate) at that frequency. This interaction could stimulate plasma wave growth and produce an amplification of plasma waves in the medium. If the growing plasma wave is then coupled to an EM resonator, this "catcher" cavity could provide a conversion from a plasma wave back to an EM wave. The result would be a plasma amplifier of EM waves. (For further description of the catcher cavity concept, see Chapter 9, Figure 9-2.)

From Fig. 7-8, it can be seen that the amplifier would operate over a band of wavelengths varying from k_1 to k_2. This band would be fixed by the plasma parameters and the drift velocity V_2 of the electrons in the plasma. The amplifier might then be able to operate over a wide bandwidth (perhaps over an octave), since the growing wave (or turbulence) bandwidths in the plasma are generally broad, as has been observed in the laboratory and also in the natural aurora.

The k-vector used in plasma theory can be described by

$$\bar{k} = \hat{k}_1 \,|\, \bar{k}\,| = \hat{k}_1 \left| \frac{\omega}{V_{ph}} \right| ,$$

where

$$\hat{k}_1 = \text{unit vector in wave direction}$$

$$\omega = \text{wave angular frequency}$$

$$v_{ph} = \text{phase velocity of plasma wave.}$$

The wave vector $\bar{k}$ comes from the turbulence theory and suggests a wave fluctuation in a fluid that propagates in a given direction $(\hat{k}_1)$. The unit vector $\hat{k}_1$ is just a vector of magnitude one with the direction of $\hat{k}_1$. The wave vector of the plasma wave must match the magnitude and direction of the wave vector of the EM wave. The wave vector of the EM wave is given simply by

$$k = \frac{\omega}{V_{ph}} = \frac{2\pi}{\lambda_0} ,$$

in free space.

This matching of wave vectors transfers momentum from the EM wave to the plasma wave. This could be done by an input cavity that is similar to that used in the kystron tube. After the plasma wave travels down the plasma tube and is amplified, the reverse process is performed, and an EM wave is coupled out of the output cavity. In the latter case, the momentum of the grown plasma wave is transferred to the EM wave at the output cavity.

An example of the plasma wave amplifier concept is given in Fig. 7-10. An electron emitter (cathode) injects electrons into the plasma, and the stream is compressed by the magnetic field B_0'. The stream enters the input cavity region, and the electric field E_0' (due to the input signal E_1) is used to stimulate normal modes in the plasma column. These waves proceed along $\hat{z}$, which coincides with the current stream velocity V_d. A longitudinal magnetic field $\overline{B}_0$ is used to collimate the plasma, reducing losses to the walls of the column. The waves flow to near saturation levels at the plane x-x′ and excite the grids of the output cavity, giving an amplified version of the input signal E_1.

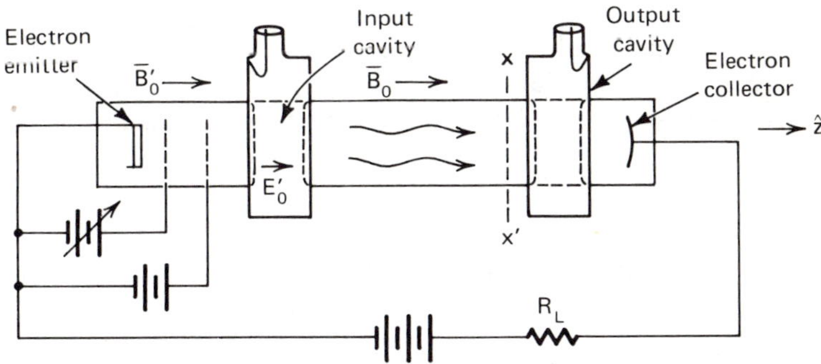

Figure 7-10: Generic Plasma Wave Amplifier
(Narrow-Band Device Concept)

This amplifier would be narrow band due to the use of tuned input and output cavities. The normal modes of the plasma would be chosen to give maximum growth at the appropriate k-vector value. The magnetic field values in this case do not tune the amplifier: they are collimating fields, not resonating fields.

There are a couple of similarities between this type of device and the space charge wave devices in Chapter 9: both use electron charge fluctuations that grow in space as they travel in a preferred direction; and both require some sort of grid or cavity structure to couple EM wave effects into the active, or amplifying, medium. In the plasma amplifier, however, the space charge of the electrons is neutralized by the positive ions in the plasma. This natural property of neutralization of space charge, may permit higher output powers from the device. The gyro-monotron to be discussed in Chapter 12, for example, was found to produce about six times more peak power with a plasma present in the interaction space than without. The plasma in this case was not used to stimulate waves, but only to neutralize the space charge of the hollow electron beam of that device. The plasma frequency was considerably less than the gyro-monotron oscillator's operating frequency, giving little possibility of plasma waves interactions. However, the plasma raised the peak output power of the device to 60 MW at 9 GHz.

Using a plasma as an amplifying medium has certain advantages. The first of these is the neutralization of coulomb forces, permitting higher beam current densities in the devices. The second is the "self-healing" properties of the plasma. Local breakdowns in this medium are of no great consequence as new plasmas tend to form continuously. Both of these properties favor high-power operation. A third advantage is that the resonant frequency is set by the plasma frequency, and not the resonator cavity frequency. In some cases, the plasma frequency is set by the discharge current in the plasma, so that and external current can vary the plasma frequency in the device. It can also be used to bring about the desired variation in velocity distribution functions, as mentioned previously.

7–8 INSIGHTS INTO THE FUTURE

The principles of plasma wave-EM interaction and the generation of plasma waves have been discussed. Plasmas and plasma waves occur in nature and thus are not new phenomena. Our understanding of them is recent, however.

Plasma wave amplifiers and oscillations have not been developed as of this writing, but they may be developed as our understanding of the mechanisms in the plasma improves. Such understanding promises development of very high power oscillators and amplifiers, based upon the principles discussed in this chapter. They could be developed using some of the same stimulation processes known to exist in the plasmas of the aurora borealis and the aurora australis. Research in the mechanisms of the auroral plasmas may provide new mechanisms for plasma amplifier devices.

8 M-type Crossed-Field Devices

8-1 INTRODUCTION

An important class of microwave devices is the M-type vacuum oscillator and amplifier. The term *M-type* comes from the French *TPOM* *(tubes à propagation des ondes à champs magnetique)*, which literally means 'tubes for the propagation of waves in a magnetic field.' The M-type device is the basic crossed-field device, i.e., a device in which the internal electron stream is influenced by crossed electric and magnetic fields. (The category of the O-type device, the one involving a linear electron beam and parallel electric and magnetic fields is covered in Chapter 9.) The electron stream in the M-type device is not generally a beam but can be a radially expanding stream of electrons. In general, the electron moves in a direction that crosses the magnetic field, giving a force on the electron that causes it to move in a curved path. This effect is sometimes called the *magnetron principle*.

There are many types of crossed-field devices, and not all of these will be discussed in detail. Table 8-1 is a tabulation of the M-type devices. The oldest and most important of these is the magnetron tube. Since many of these devices have common operating principles, this chapter will concentrate on the common electronic mechanisms involved in all M-type devices, so that the general principles can be grasped.

After the discussion of M-type electronic principles, examples are given of present-day devices. This will be followed by examples of circuits required in device use. This latter will be followed in turn by examples of system applications that use the device and its coupling circuits.

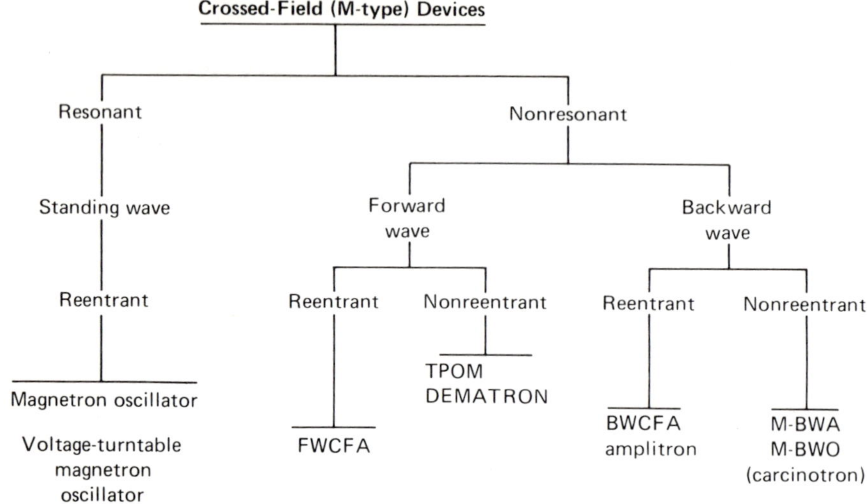

Legend: FWCFA—Forward wave cross-field amplifier
BWCFA—Backward wave crossed-field amplifier
M-BWA—M-type backward wave amplifier
M-BWO—M-type backward wave oscillator
TPOM—Injected beam CFA

Table 8-1: Representative Crossed-Field (M-Type) Microwave Devices

8–2 M-TYPE DEVICE PRINCIPLES

In the M-type device a dc electric field is used to accelerate the electron, and a perpendicular magnetic field is used to cause the electron to move in a curved path. The path of the electron in the device is not linear, due to the force of deflection provided by the magnetic field, but an rf interaction is made possible by this process.

The magnetron principle is illustrated in Fig. 8-1, where the magnetron action occurs between two parallel plates.

The electric field $\mathcal{E}_o$ created by the battery potential across the two plates will move the electron in the y-direction due to its negative charge. The velocity given the electron by this force will move it across the magnetic field B_0. (The magnetic field is assumed to fill the area between the plates uniformly.) The force on the electron due to the electron motion across the magnetic field direction is in a direction perpendicular to E_o and the electron velocity. The paths of the electron as a function of the field strength B_0 are shown in Fig. 8-2.

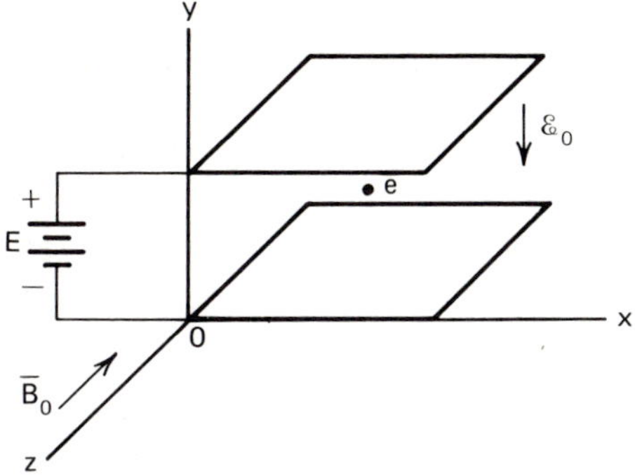

Figure 8-1: The Planar Magnetron Principle

For a small magnetic field, path a will be followed. The electron will be only slightly deflected. A stronger magnetic field will cause the electron to graze the anode and return to the cathode, as in path b. A very strong value of B_0 will cause the electron to move in the cycloidal path c. The motion of the electron in the y-direction will be periodically away from the cathode and back at a frequency f_c which is proportioned to the strength of the field B_0. In practical units,

$$f_c = 2.8 \times 10^6 \, B, \qquad (8.1)$$

where B is in gauss. The theoretical values of the field B_0 and the anode voltage V_2 for grazing electron paths for the cyclindrical and planar magnetron cases are given in Table 8-2.

Consider now an alternating field added in series with the battery E in Fig. 8-2. Set the frequency of the alternating field to f_c. This frequency can be expected to couple with the electron's motion. Assume now that an electron leaves the cathode at an instant in the alternating field cycle that accelerates the electron in flight and then pushes it toward the cathode in its return flight (path c of Fig. 8-2). In this case the electron receives energy from the ac field and so moves with a higher velocity. If the

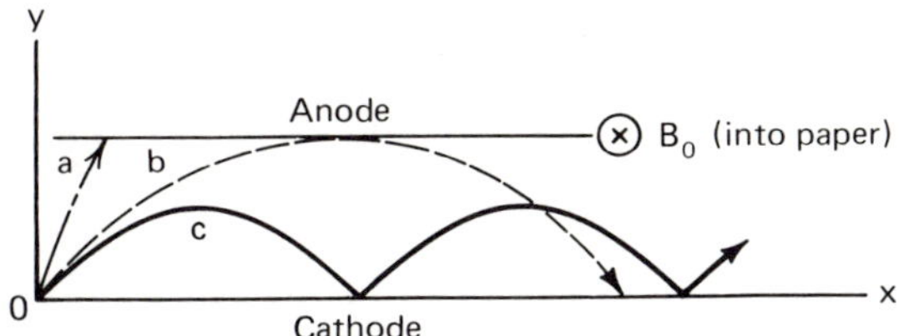

Figure 8-2: Electron Paths in a Planar Magnetron

opposite occurs, and the ac electric field opposes the cycloidal motion of the electron over its path, then the electron loses velocity (energy). These two cases are illustrated in Fig. 8-3.

The mechanism, then, is that an electron in crossed fields can either absorb energy or give energy to the alternating plate voltage source. In a magnetron oscillator the electrons that deliver energy to the anode lose velocity but increase the amplitude of the ac field at the anode. The electrons that leave the cathode later in the rf cycle and receive energy from the ac field are given higher velocity (energy) and are driven into the cathode, causing heating of the cathode surface. It can be shown that the electrons that give energy to the ac field make many more trips than those that absorb energy; thus, there is a net delivery of energy from the electrons to the source of the ac field. It can also be shown that the frequency of the ac generator in series with E of Fig. 8-1 must be nearly equal to f_c for maximum energy transfer.

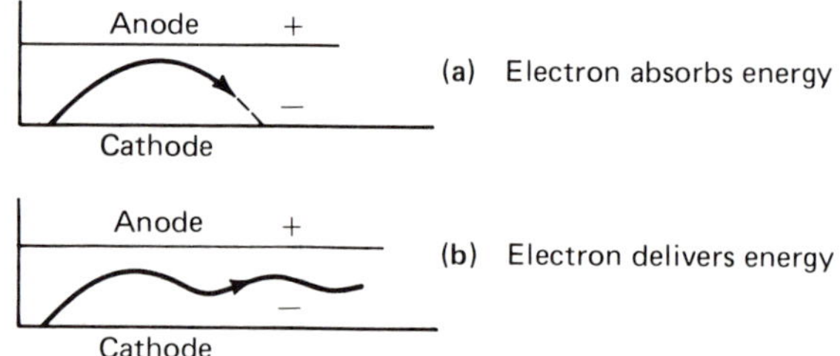

Figure 8-3: Electron in Magnetron with AC Field on Anode

Figure 8-4 shows the elements of a planar magnetron oscillator. The resonant circuit (L-C-R in parallel) is tuned to f_0, and appropriate values of E and B_0 are applied. f_0 is the value corresponding to the resonant circuits, and is given by

$$f_0 = \frac{1}{2\pi \sqrt{LC}} = f_0 \qquad (8.2)$$

If the losses represented by the value of R match the gain due to electron interaction with the anode field, then sustained oscillations will occur. A source of electrons at the cathode, together with the anode electric field, is necessary.

Analysis shows that the electrons cannot remain too long in the interaction space because of a gradual change in oscillation phase that occurs with time. In most cases, however, the finite size of the electrodes in a planar magnetron make it unnecessary to have any special design for removal of these electrons.

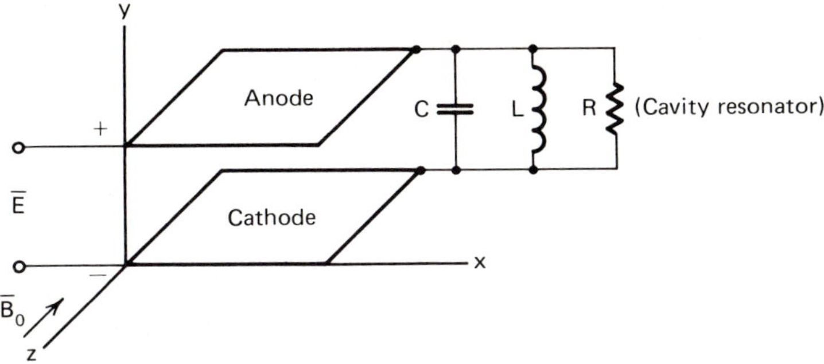

Figure 8-4: The Planar Magnetron

8–3 TYPES OF MAGNETRONS

There are three basic types of practical magnetrons:

(1) Negative-resistance magnetron

(2) Cyclotron frequency magnetron

(3) Travelling wave magnetron.

The negative-resistance magnetron uses a static negative resistance between two anode segments to sustain oscillations, but it is usually operated at frequencies lower that that of the microwave region. The cyclotron frequency magnetron can use a cylindrical configuration or a split anode geometry, as illustrated by Fig. 8-5.

The planar magnetron of Fig. 8-4 is not practical, as it requires an excessively long interaction space to achieve reasonable energy gains. Therefore, only the above-mentioned three magnetron types are used in

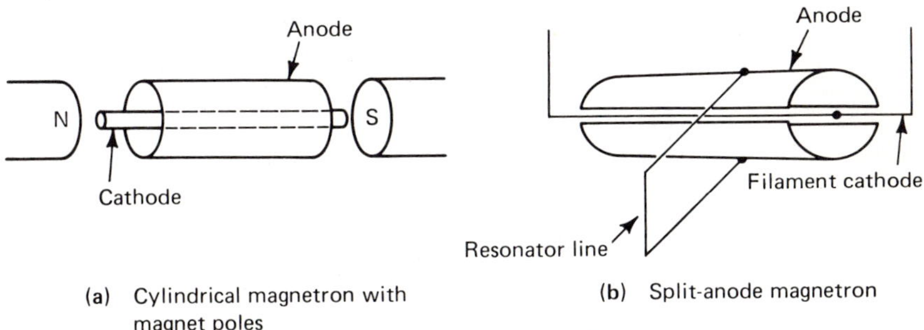

Figure 8-5: Cyclotron Frequency Magnetron Designs

practice. The above analysis for the planar geometry magnetron, however, serves to illustrate the basic magnetron interaction principle.

For the cyclindrical cyclotron frequency magnetron of Fig. 8-5a, the frequency of the radial component of the electron motion is related to the cyclotron frequency by

$$f_c' = f_c/\sqrt{2} = 2 \times 10^6 \, B,$$

where B is in gauss. The theoretical values of the field B_0 and the anode voltage V_2 for grazing electron paths for the cyclindrical and planar magnetron cases are given in Table 8-2.

CASE	CYLINDRICAL MAGNETRON	PLANAR MAGNETRON
Expression	$r_a^2 B^2 = 4.55 \times 10^{-7} V_a$	$s^2 B^2 = 1.14 \times 10^{-7} V_a$
Units	B, gauss r_a, cm (cylinder radius) V_a, voits	B, gauss S, cm (plate separation) V_a, voits

Table 8-2: Magnetron Resonance Conditions

The conditions for magnetron resonance for the two geometries of Table 8-2 are different because of the two different anode-cathode geometries. Otherwise, the principles of operation are very similar.

The travelling-wave magnetron is the one most used today at microwave frequencies, due to the simplicity of its operation at its efficiency level. In this type the geometry and electron path are as shown in Fig. 8-6.

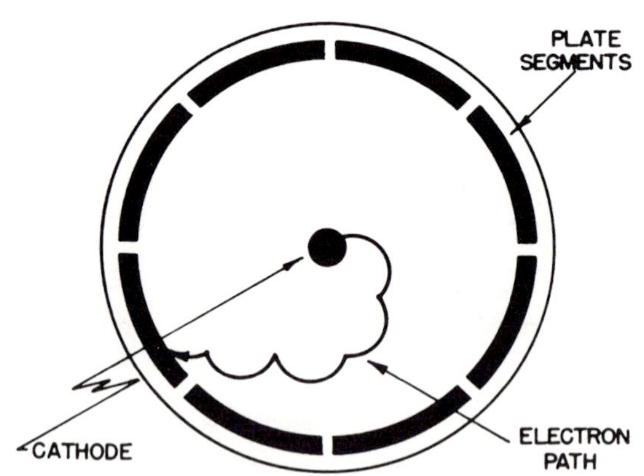

Figure 8-6: Electron Path in Travelling-Wave Magnetron

The probable path of electrons for a practical design whose resonant cavities are placed in the anode structure is given in Fig. 8-7. Note that for this oscillation mode, alternate anode segments alternate in polarity. In Fig. 8-7 electron Q_1 receives energy from the rf field and is thus promptly returned to the cathode. Electron Q_2 gives energy to the rf field and eventually is collected by the anode. This action is similar to the planar case. (See Fig. 8-3.) The difference now is that the structure is reentrant; that is, the electron can reenter its previous path. The reentrant feature is what permits a travelling-wave mode.

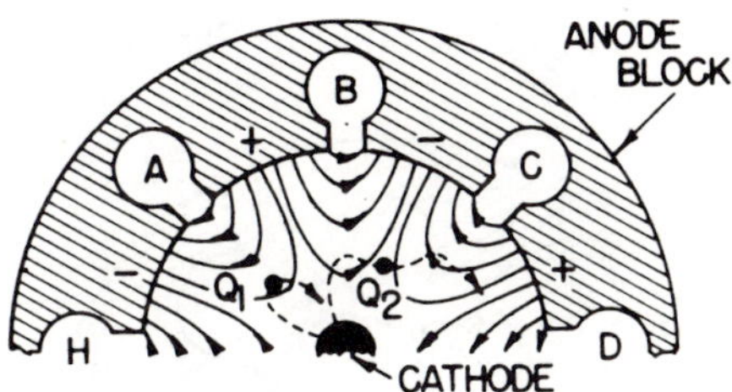

Figure 8-7: Travelling-Wave Magnetron with Electron Paths during Oscillation.

Electrons such as Q_2 in Fig. 8-7 are called working electrons, since they stay in the interaction space for considerable time before striking the anode. The cumulative action of the electrons being returned to the cathode and being directed toward the anode results in a pattern similar to the spokes of a wheel, as indicated in Fig. 8-8.

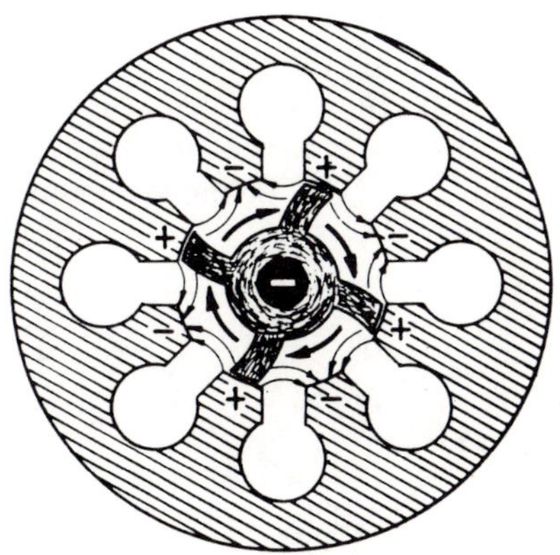

Figure 8-8: Rotating Space Charge Wheel in an Eight-Cavity Magnetron

Electrons emitted from the cathode between wheel spokes are quickly returned to the cathode, since they are not working electrons. The mode illustrated in Fig. 8-8 is called the *Pi-mode*, since each segment differs by π radians, or 180°, in phase. The Pi-mode is the preferred mode, since it is not degenerate. All the other possible modes are degenerate, i.e., they can oscillate at the same two mode frequencies but with an interchange in nodes and antinodes in the standing wave patterns. To discourage other possible resonant modes, straps are sometimes connected between anode segments of the same phase. This also moves other modes to frequencies higher than the Pi-mode, giving better mode separation. The strapping case is illustrated in Fig. 8-9.

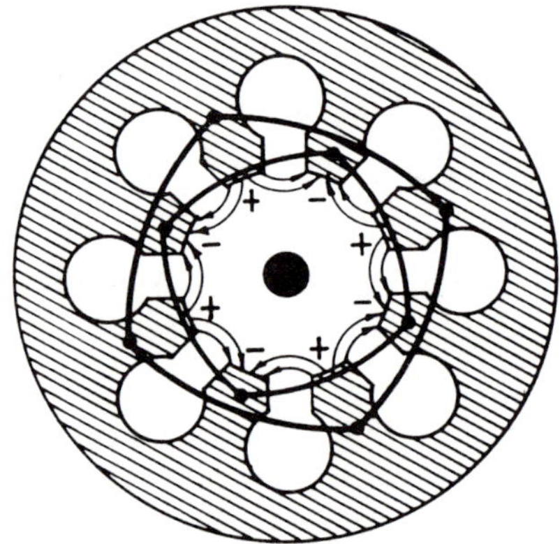

Figure 8-9: Alternate Segments Connected by Strapping Rings

For the travelling wave magnetron, each cavity, together with its slot, forms a resonant circuit, as illustrated by Fig. 8-10. In the Pi-mode the cavity resonators are all effectively in parallel, in spite of the circuit of Fig. 8-11, which has the appearance of a series connection. But the parallel connection is the true one, because in the Pi-mode all the plus

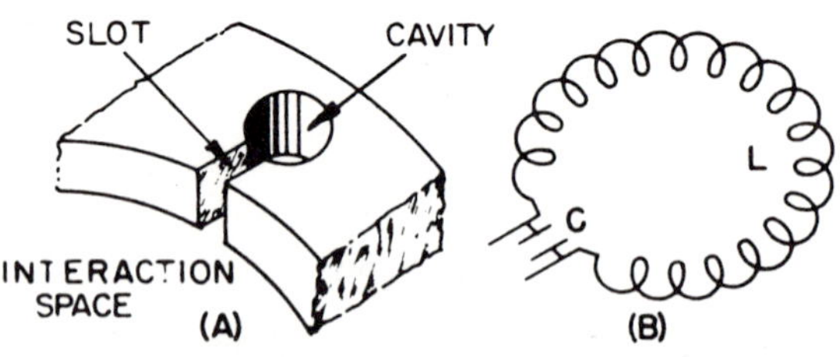

Figure 8-10: Equivalent Circuit of a Hole-and-Slot Cavity

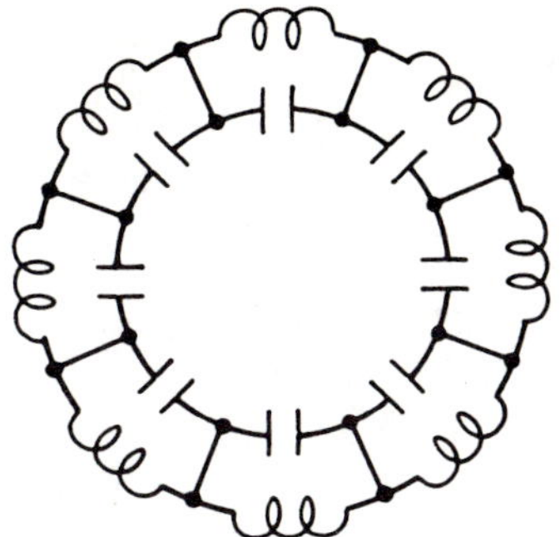

Figure 8-11: Cavities Connected

terminals of Fig. 8-11 are in phase. They may all, then, be considered to be strapped together by ideal straps. So for N cavities, the total capacitance is NC, and the total inductance becomes L/N. The total LC product is then $(NC) \cdot \left(\dfrac{L}{N}\right) = LC$, or the LC product of one resonator, and the resonant frequency is that of a single cavity. (See Equation 8.2.)

Some common types of anode blocks are given in Fig. 8-12. The rising sun block has a frequency somewhere between that of the large slot and small slot resonators.

Several methods have been developed for tuning magnetrons via the variation of the cavity resonances. Figure 8-13 shows inductive tuning via a "sprocket," or "crown of thorns" tuner. Figure 8-14 illustrates capacitive tuning via a "cookie cutter" tuner. Either of these tuners can give a 10% tuning range for the magnetron. The two tuning methods could be used in combination to cover an even greater tuning range.

8–4 THE MAGNETRON "BAKING-IN" PROCEDURE

During the initial operation of a high power magnetron, internal arcing is common. Gas is normally liberated from tube elements during idle periods, and this contributes to arcing. This condition is common in new tubes also. In either case, the condition can be relieved by employing a fairly simple "baking-in" procedure. Anode voltage is gradually applied, until arcing occurs at a rate of several times a second. The voltage is held at this level until arcing dies out. The voltage is raised again until arcing occurs at the above given rate and is held until arcing again dies out. The

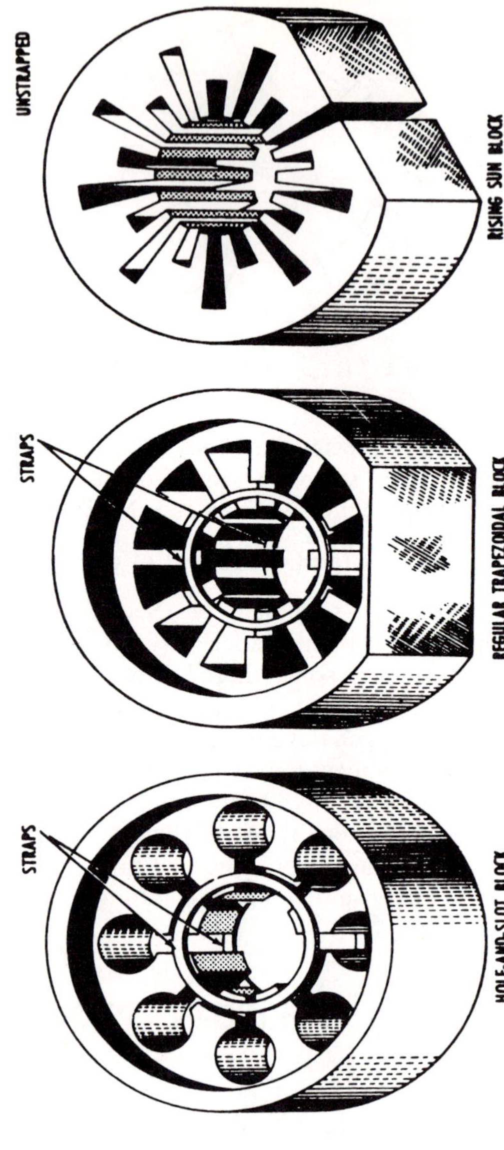

Figure 8-12. Common Types of Anode Blocks

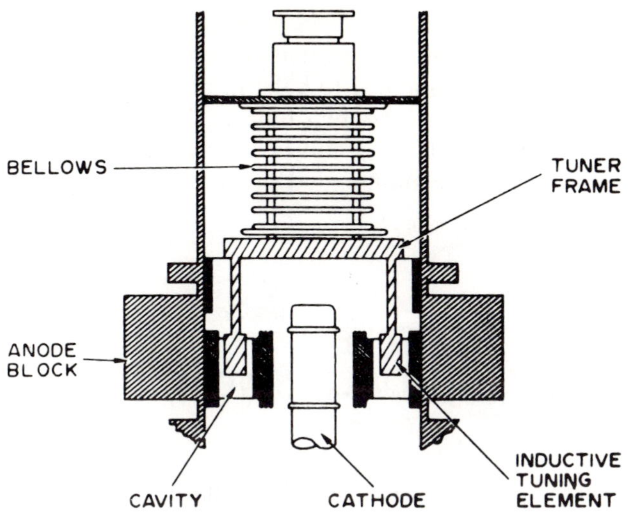

Figure 8-13. Inductive Magnetron Tuning

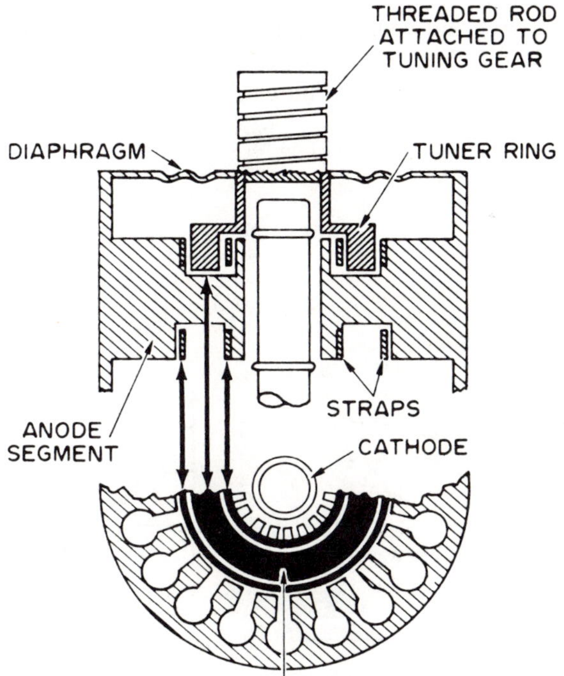

Figure 8-14: Capacitive Magnetron Tuning

procedure is continued, and when the normal rated voltage is achieved, and the magnetron is stable at the rated current, the baking-in process is complete. It is good maintenance to bake in magnetrons when they have been left idle in equipment or when long periods of nonoperating time have accumulated on individual tubes. Equipment technical manuals should be consulted for recommended times and procedures for specific magnetron types.

8–5 THE FORWARD-WAVE CROSSED-FIELD AMPLIFIER (FWCFA)

From Table 8-1, the crossed-field amplifier (CFA) can exist as either a forward-wave or backward-wave amplifier. It can also be classified according to whether it uses an electron stream source that is an injected beam or an emitting sole.

The difference between a forward-wave or a backward-wave rf circuit is the difference between the phase velocity and group velocity directions. Figure 8-15 illustrates the ω-β diagrams of the two types of wave circuits. The definition of phase velocity is the expression

$$\mathcal{V}_{ph} = \frac{\omega}{\beta},$$

where

$$\omega = 2\pi f_0,$$

and

$$\beta = 2\pi/\lambda_0.$$

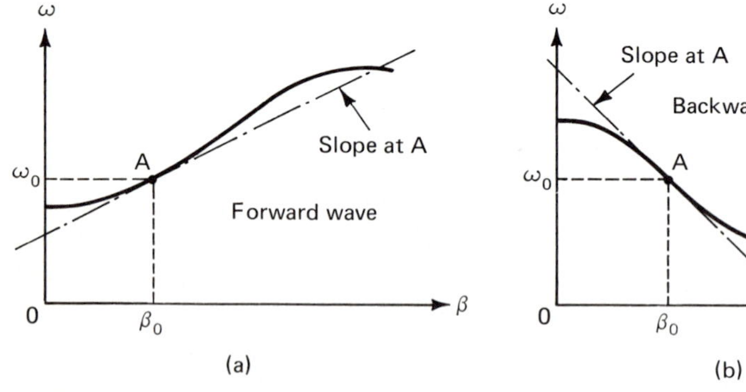

Figure 8-15: ω-β Diagrams

where f_0 is the frequency of the wave and λ_0 is its wavelength. The group velocity of the wave is given by $U_{gp} = \dfrac{\Delta\omega}{\Delta\beta}$, or the local value of the slope at a point on the ω–β curve. The energy information on the wave travels at the group velocity, but the phase information in the wave travels at phase velocity $\mathcal{V}_{ph}$. Note how, in Fig. 8-15a the point A has a local slope that is in the same direction as the ratio $\dfrac{\omega_0}{\beta_0}$ at that point. Now the local slope is the group velocity U_{gp}, and the ratio $\dfrac{\omega_0}{\beta_0}$ is the phase velocity $\mathcal{V}_{ph}$. Since both U_{gp} and $\mathcal{V}_{ph}$ are positive (uphill) in direction, the wave in the circuit illustrated by the ω-β diagram of Fig. 8-15a is a forward wave. The ω-β response of Fig. 8-15b is clearly different. At point A on the ω-β curve the ratio ω_0/β_0 is positive, but the local slope is negative (downhill). From the definitions, it is clear that $\mathcal{V}_{ph}$ is positive, but U_{gp} is negative. This means that the phase of the wave advances in the positive direction, but energy flow in the wave is in the opposite (negative) direction. This results in what is called a backward wave. The difference between a forward-wave amplifier and a backward-wave amplifier lies in the type of travelling wave rf circuit used to propagate the energy in the device.

Figure 8-16a illustrates the two types of CFA. For a given circuit energy flow direction (counterclockwise, or ccw, for example), a forward wave CFA (FWCFA) would have the electrons flow ccw along with the phase velocity, permitting wave coupling and amplification. The delay line would then have the ω-β diagram of Fig. 8-15a. For a backward wave CFA (BWCFA), the electron flow and phase velocity of the wave would be clockwise (cw), and the circuit energy flow would be ccw. This would require the ω-β response of Fig. 8-15b for proper coupling of the electron stream to the phase velocity of the delay line circuit.

Figure 8-16a uses a continuous emitting-sole cathode, and Fig. 8-16b uses an injected electron beam. The sole plate is a special electrode used in CFAs. Thus, Fig. 8-16 illustrates FWCFA and BWCFA devices, and also the injected-beam and emitting-sole devices. All have a magnetic field crossed to the electric field, qualifying them as M-type devices.

8–6 THE M-CARCINOTRON (M-BWO) DEVICE

The M-carcinotron oscillator is an M-type backward-wave oscillator, as it uses crossed electric and magnetic fields in the interaction space. A

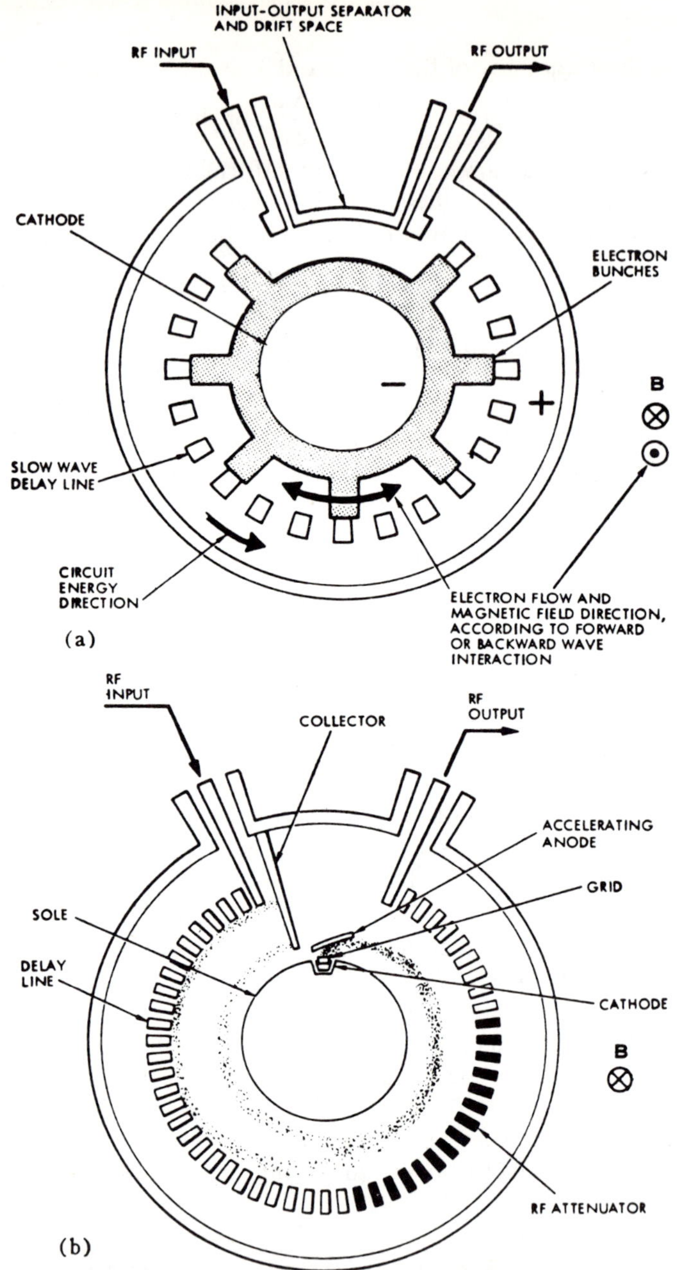

Figure 8-16: Schematic Diagrams of CFAs (a) Continuous Emitting-Sole Cathode CFA, Forward Wave or Backward Wave. (b) Injected-Beam CFA. *

*J.F. Skowron, *The Continuous Cathode (Emitting Sole) Crossed Field Amplifier*, Proceedings IEEE, Vol. 61 #3, pg. 330–356, March 1973. 1973 by IEEE. Reprinted by permission of IEEE.

linear model of an M-carcinotron is illustrated in Fig. 8-17. Electrons are drawn from the cathode surface by the accelerating potential V_0 and bent into a curved path by B. The electron beam is injected parallel to the sole and flowing toward the collector plate. The effects of E and B balance, resulting in a parallel electron flow past the slow-wave structure. The slow-wave structure is backward-wave, so that oscillation builds up in a right-to-left direction, with the maximum wave amplitude at the left end of the structure. An amplifier can also be designed using the same principles, with the rf termination of Fig. 8-17 replaced by the amplifier's input terminal.

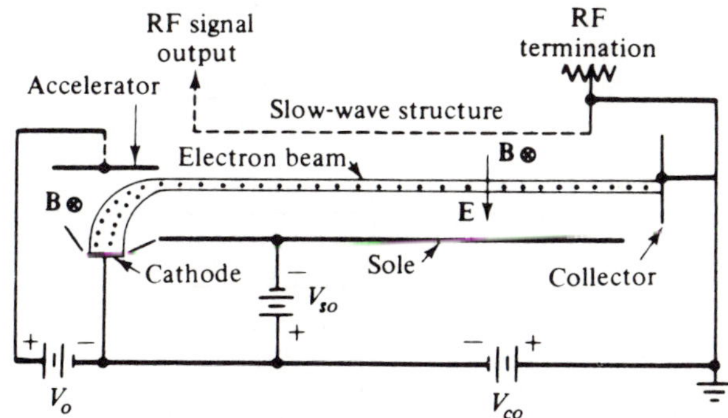

Figure 8-17: Linear Model of an *M*-Carcinotron oscillator *

8–7 APPLICATIONS OF M-TYPE DEVICES

All M-type devices operate so that the active electronic region of the device has electric and magnetic fields oriented perpendicular to each other. The M-type devices are capable of very high peak power outputs, making them quite suitable for use in high-power radars. The state of the art for U. S. high-power magnetrons is depicted in Fig. 8-18. Peak power levels of 10 MW (10^7 W) have been achieved at about 1 GHZ. One magnetron achieved a pulse level of 40 MW at 10 GHZ, a point that is off the graph.

---

*J.V. Gewartowski and H.A. Watson, Principles of Electron Tubes, pg. 391, D. Van Nostrand Co., Princeton, NJ, 1966. Reprinted by permission of Wadsworth Publishing Co., Belmont, California.

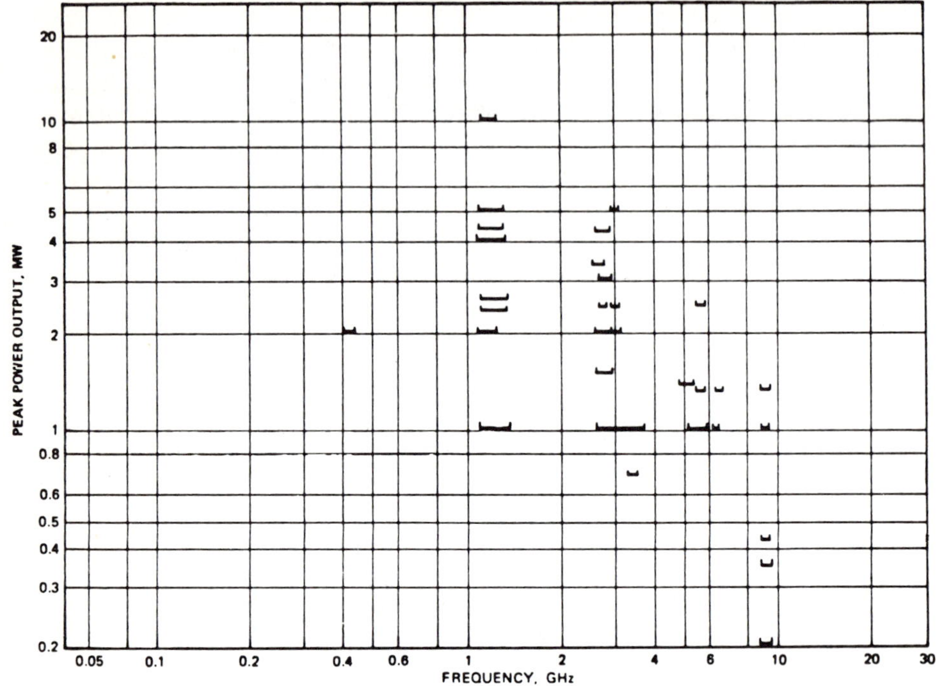

Figure 8-18: State of the Art for U.S. High-Power Magnetrons *

The state of the art for U. S. crossed-field amplifiers (CFA) is depicted in Fig. 8-19. This figure includes the forward-wave crossed-field amplifier (FWCFA), the dematron, the amplitron, and the carcinotron.

The M-type oscillator and amplifier achieves its high-power levels at high efficiency. In general, the plate supply voltage for the crossed-field tube is less than that required for the linear-beam tube. 40 KV is typically required for a 1-MW pulse output. Efficiencies for a circular, reentrant design can vary from 30 to 70%, with 45% being typical. The usable dynamic range is less than that of the linear-beam tube, varying from 3 to 20 dB. The tube is also smaller in size and weight compared to the linear-beam equivalent. But the crossed-field tube tends to generate more spurious noise than the linear tube, and thus, such designs are not suitable for low-power applications, such as local oscillators in microwave receivers.

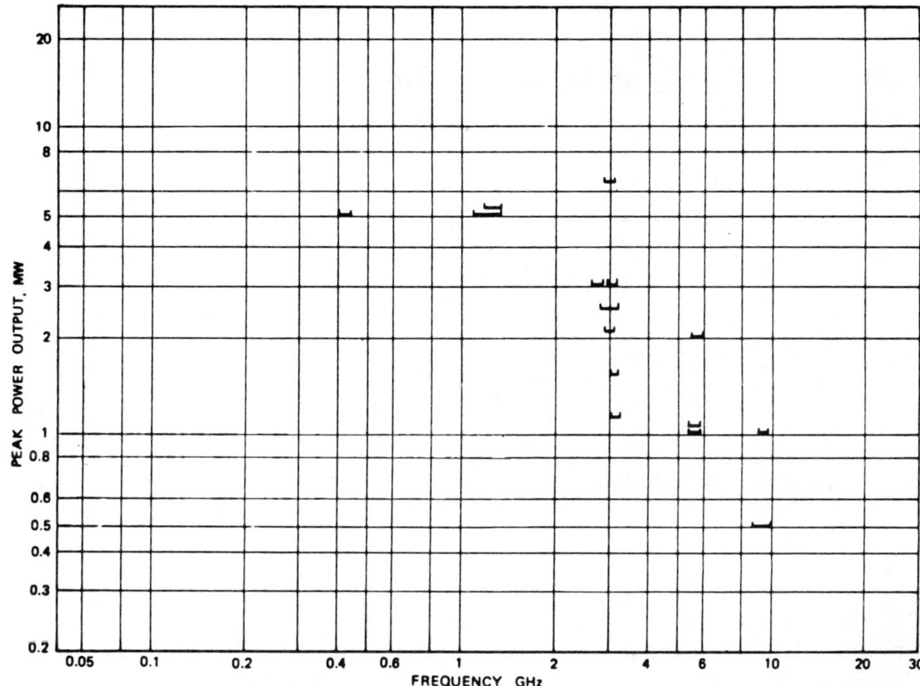

Figure 8-19: State of the Art for U.S. High-Power CFAS

A common commercial use of the magnetron is in the microwave oven. Since only a dc power supply is required to generate microwave energy from a magnetron at high efficiency, its simplicity and low cost has made it possible to be used as a household item.

---

*Samuel Y. Liao, Microwave Devices and Circuits, ©1980, 243. Reprinted by permission of Prentice-Hall, Inc., Englewood Cliffs, NJ.

9 O-Type Parallel-Field Devices

9–1 INTRODUCTION

The previous chapter considered the M-type microwave device and its characteristics. This chapter will consider the O-type device mentioned previously. The term *O-type* comes from the French *TPO* (*tubes à propagation des ondes*), which literally means 'tubes for propagation of waves.' This class of tubes is also called *linear-beam type*. In these tubes a magnetic field is used that is generally aligned with the parallel electron stream and the accelerating electric field. The magnetic field tends to hold together (collimate) the electron stream. The electrons are generally bunched to permit an interaction with the output structure (cavity, helix, etc.). O-type tubes can be amplifiers as well as oscillators.

There are many types of linear-beam (O-type) devices, and not all of them will be described in detail. Table 9-1 is a tabulation of the prominent O-type devices.

9–2 O-TYPE DEVICE PRINCIPLES

In the O-type device a dc electric field is used to accelerate a stream of electrons and give them potential energy before they enter the microwave interaction region. A parallel magnetic field $\bar{B}_0$ is used to force the electron stream into a well-defined beam while the electrons travel down the stream direction. Figure 9-1 illustrates the basic linear-beam principles. The accelerating voltage V_0 gives the electrons emitted by the cathode an initial potential energy. These electrons are collimated by the

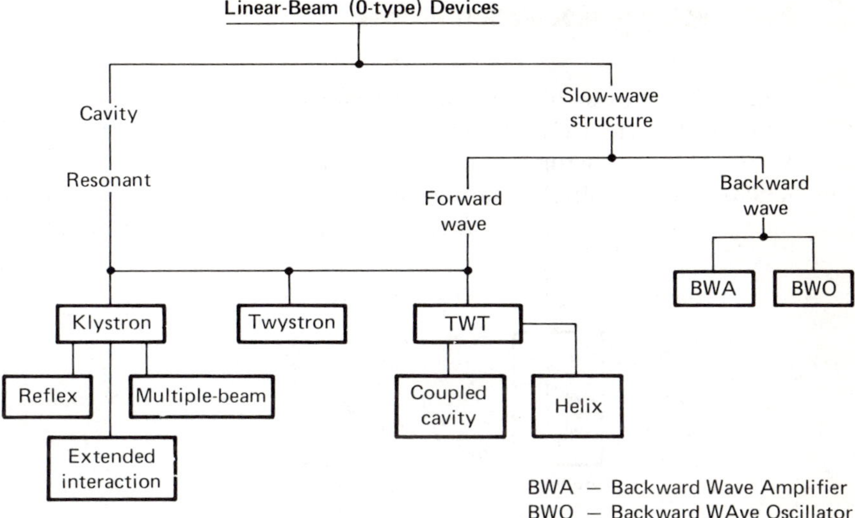

Table 9-1: Representative Linear-Beam (O-Type) Microwave Devices

anode with the hole and travel through the microwave circuit. The magnetic field $\bar{B}_0$ forces the electron stream into a well-formed beam of electrons. This beam is sent into the microwave circuit, where it is modulated in some way. As the electrons travel down the microwave circuit, they exhange energy with it, and energy is gained from the electron stream. This results in an amplification of the initial microwave electrical signal (in the case of an amplifier), or the generation of a microwave signal (in the case of an oscillator). The differences in the types of linear-beam devices, as given in Table 9-1, are due to the forms of electron stream microwave circuitry interaction.

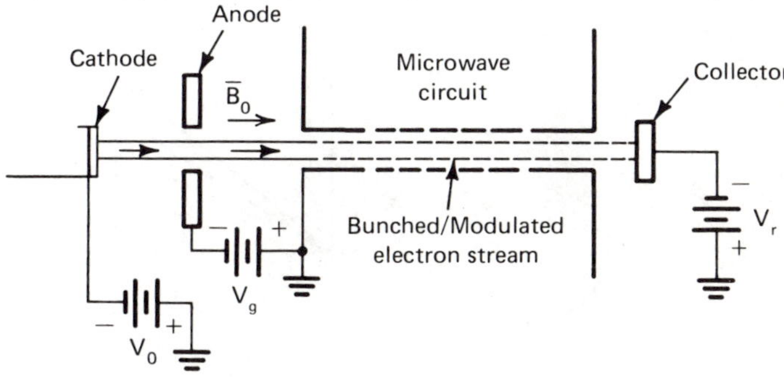

Figure 9-1: Linear-Beam Principles

9–2.1 Velocity Modulation Process

Figure 9-2 shows the type of microwave circuit used for velocity modulation of the electron stream and bunching of the electrons. In this case V_r is set to zero for full electron collection. The rf input to the buncher cavity causes an alternating electric field in the z-direction (across G_1) that will alternately accelerate and decelerate the electron stream,

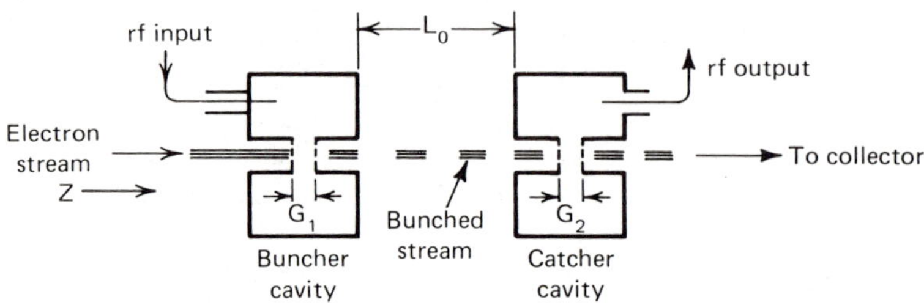

Figure 9-2: Microwave Circuit/Beam Interaction for Velocity Modulation

depending on when in the rf cycle the electron enters the grid space. Figure 9-3a shows the voltage across G_1 of Figure 9-2. At point a in the ac cycle the electron entering the grid area will be accelerated momentarily and will have additional velocity when it leaves the grid area. At point b there is no voltage, so no added velocity results. At point c in the rf cycle the electrons entering at that instant will be momentarily decelerated and will lose velocity. (See Fig. 9-3b.) At a time T_0 seconds later the electrons will be at different distances from the buncher grid, depending upon when they entered in the rf cycle. Figure 9-3b, shows that at T_0 seconds some electrons have moved out to distance l_a due to the acceleration at point a; some are at some mean distance l_b; and some will be

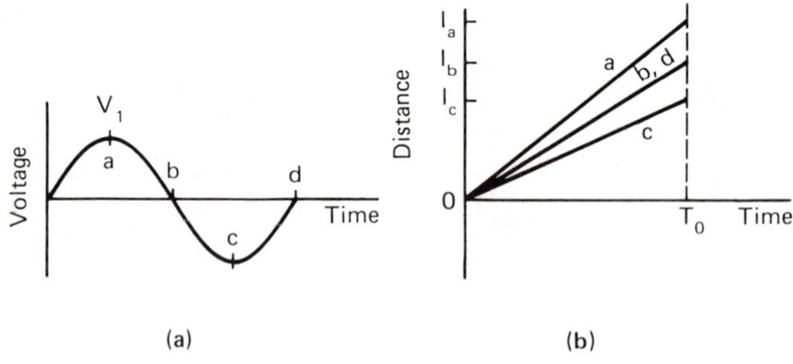

Figure 9-3: Velocity Modulation

only at distance l_c due to the deceleration at point c. At some greater distance downstream the higher-velocity electrons will catch up to the lower-velocity electrons from a previous rf cycle, and the electron stream will be made up of bunches of electrons. The process is illustrated by the Applegate Diagram of Fig. 9-4. In the figure, the shaded areas represent those regions in which the electrons catch up to form bunches. The distance between the buncher and catcher cavities is adjusted to optimize the bunching process. In this description the effect of the finite gap between the adjacent bunching grids is neglected. This finite gap gives a finite transit angle. The transit angle is the number of electrical degrees of the rf cycle during which the electron is between the two buncher grids.

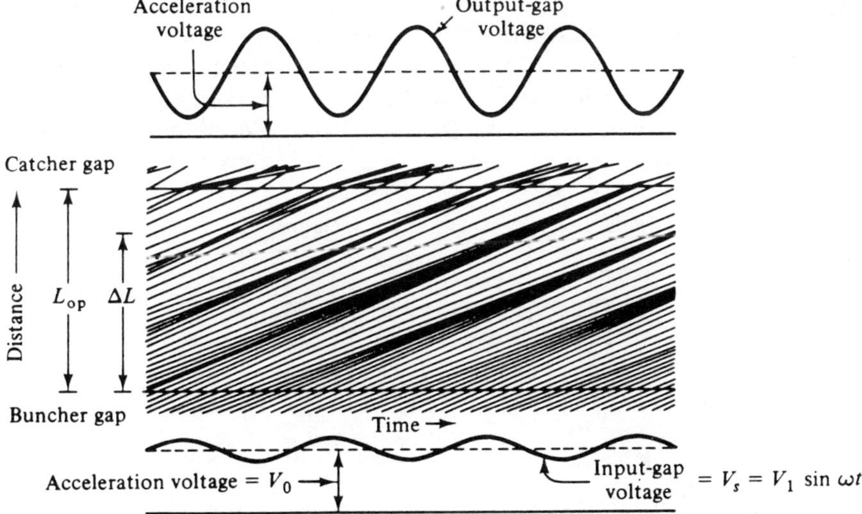

Figure 9-4: Applegate Diagram *

When the bunched electron stream reaches the region of the catcher cavity grids, the electron bunches excite the catcher cavity, and an amplified signal is induced there. The process then brought about by the ac electric field appearing between the buncher cavity grids is the modulation of the velocity of the electron stream passing between these grids, so that downstream from these grids the electron stream will consist of bunches (clumps) of electrons. Such bunches will induce a field in the catcher cavity. The process is one of amplification, since the energy in the electron stream is much greater than that in the input (buncher) cavity. Modulating a small portion of the energy (velocity) of the electron stream will result in a power gain through the tube. There is an analogy here with the conventional vacuum triode tube. The grid controls the electron stream flow, as in the conventional vacuum tube. However, the transit-time effects are deliberately used in the velocity modulation tube, whereas in the conventional triode the transit time effect was detrimental and set an upper limit on the frequency of operation of the triode. The velocity modulation process is the key to the operation of the kylstron amplifier and oscillator.

9–2.2 The Continuous Interaction Modulation Process

If the buncher cavity of Fig. 9-2 is replaced by another rf structure (such as a wire helix) that permits a continuous interaction between the electron stream and the field of the helix, then another type of amplifier is possible. Figure 9-5 illustrates this case. A helix will slow down the injected electromagnetic wave so that its phase velocity can match that of the electron stream as it travels down the tube axis.

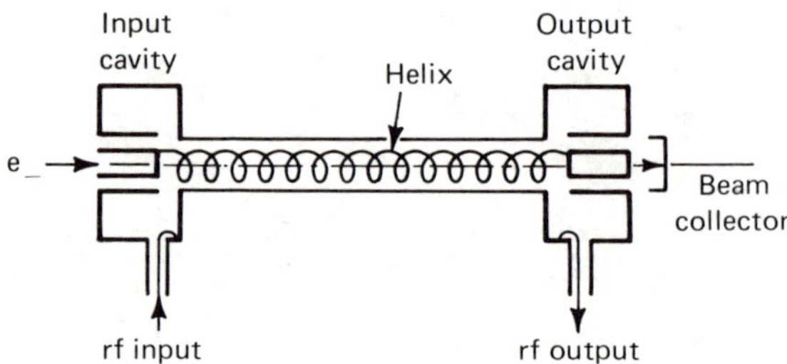

Figure 9-5: Helix-Type Travelling-Wave Interaction Tube

The energy interchange in this kind of amplifier is illustrated by the case of an electron (e_-) being trapped in a travelling potential well.

Consider now Fig. 9-6. An electron is moving at a velocity v_0 near a potential well of depth V_0 that is travelling with a velocity v_{ph}, the well's phase velocity. Now if v_0 exceeds v_{ph} when the electron is trapped by the

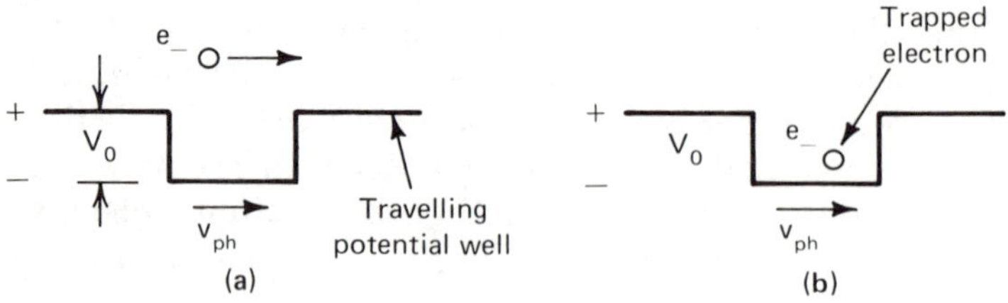

Figure 9-6: The Potential Well Analogy

well, the electron has lost velocity and thus some energy. This lost energy is actually given to the travelling potential well V_0. If v_0 is less than v_{ph}, on the other hand, the electron will receive energy from the potential well, and the energy flow will be from well to electron. Thus, if the electron stream is adjusted to have a velocity v_0 somewhat greater than that of the phase velocity v_{ph} of the travelling wave, then, on an average, the electron stream could feed energy to the electromagnetic wave in the helix. The continuous interaction for the case of Fig. 9-5 would then result in wave amplification as a signal travelled down the helix of the tube. This is the principle of the travelling-wave O-type amplifier tube.

9–3 TRAVELLING-WAVE TUBE MECHANISMS

The wire helix of Fig. 9-5 can be replaced by a series of coupled cavities that will have a similar effect. Figure 9-7 gives an example of a coupled-cavity TWT design. The cavities of Fig. 9-7a are coupled via the cavity irises (round holes). The waves are slowed down as they propagate between cavities (see Fig. 9-7b), similarly to the helix case, and the electric fields move from the gaps a-a′, b-b′, and c-c′ of Fig. 9-7 with a reduced phase velocity that nearly matches the velocity of the electron beam. Using the potential well analogy of Fig. 9-6, we can see that the electron beam velocity should be slightly greater, on the average, than the phase velocity of the coupled-cavity system. Then, the electron beam will feed energy into the coupled-cavity system and will be gradually bunched with increasing intensity as the waves propagate and interact as they travel down the beam axis. In addition, it can be shown that in the travelling-wave mode the electrons that deliver energy to the microwave structure

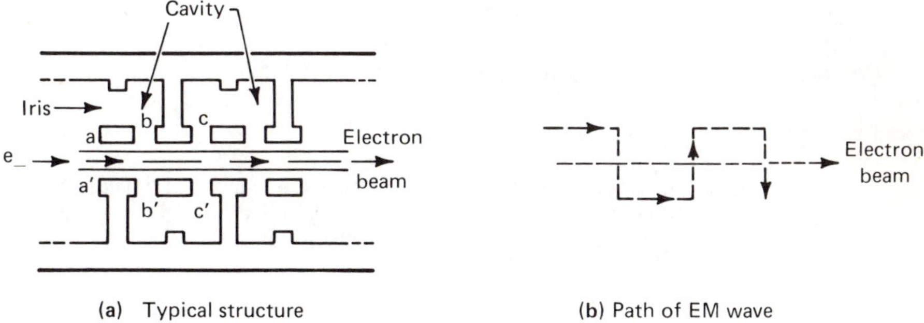

(a)   Typical structure   (b) Path of EM wave

Figure 9-7: Coupled-Cavity TWT Design

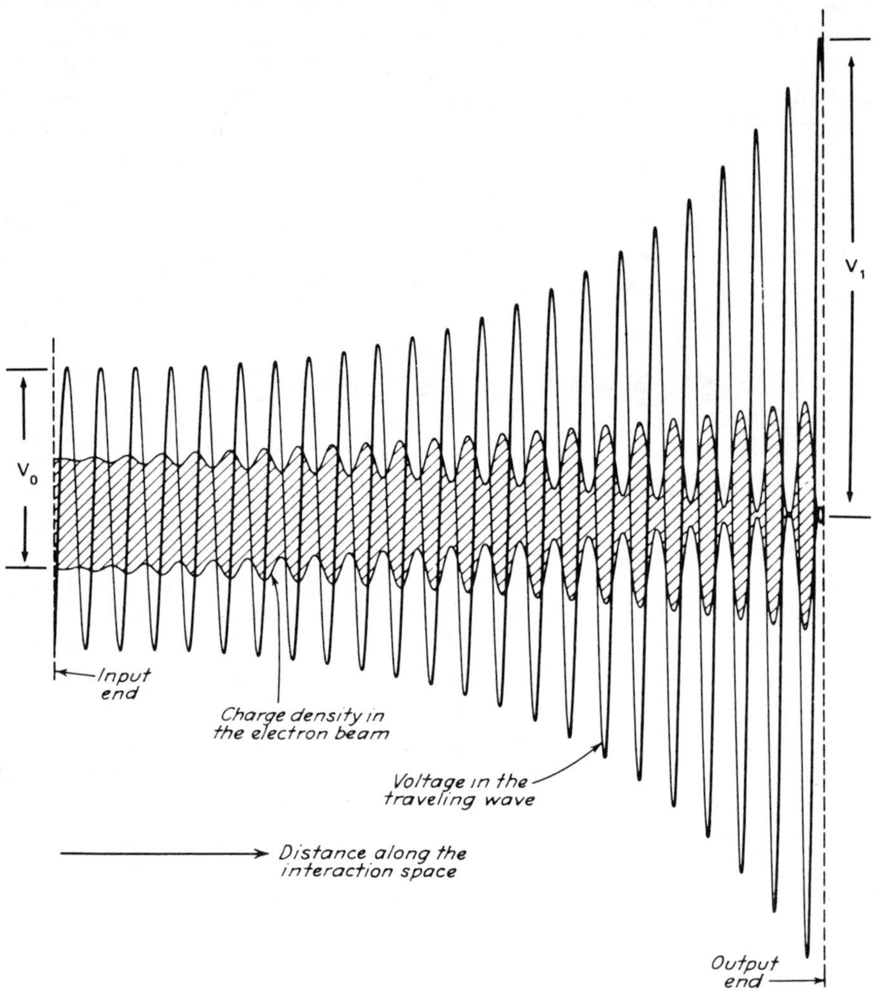

Figure 9-8: Voltage and Charge Buildup in a TWT *

---

*Reich, Ordung, Krauss, and Skolnik, Microwave Theory and Techniques, pg. 794, D. Van Nostrand Co., Princeton, NJ, 1953. Reprinted by permission of Wadsworth Publishing Co., Belmont, California.

(helix or coupled cavity) tend to lose velocity (energy) more or less continually along their complete trajectory, similar to what occurs under the magnetron action of Fig. 8-3. Thus, if the travelling-wave tube is properly proportioned (for phase velocity) and properly operated (in terms of electron initial acceleration), more energy will get transferred from the electron stream to the microwave structure than is lost from the structure to the electron stream, giving a net gain in the tube. The net effect of these conditions is illustrated by the computer-generated plot of Fig. 9-8. This shows how the charge density in the electron stream is gradually modulated as the stream flows toward the output end of the tube. The voltage buildup in the travelling wave follows this buildup of charged density and is also illustrated in Fig. 9-8.

The voltage buildup from amplitude V_0 to amplitude V_1 in the figure represents the voltage gain of the TWT. The computer plot is for a fixed time $t = 0$; the wave will travel from left to right as it propagates and grows in amplitude, due to the interaction with the modulated (bunched) electron stream.

9–4 MULTICAVITY KLYSTRON TUBE

Section 9–2.1 described the velocity modulation process that is basic to the understanding of the kylstron tube's amplification mechanisms. For the development of high-pulse powers and moderate cw power levels (tens of kilowatts), the multicavity klystron was developed. Figure 9-9 illustrates schematically the concept involved.

Several intermediate cavities are placed between the buncher cavity and the catcher (output) cavity. The rf input modulates the electron beam, as described in Sect. 9–2.1. The bunched beam that is formed induces an amplified rf voltage in the first intermediate cavity. This cavity is resonant to the rf frequency. The amplified voltage, in turn, modulates the electron stream with a certain phase shift, producing a more strongly bunched electron stream. The process is repeated again downstream at the second intermediate cavity, and finally, the catcher cavity receives a highly amplified wave induced into its output cavity. It is found that some detuning of intermediate cavities can increase power output. The bandwidth of the klystron can be increased by loading the intermediate cavities, but the overall gain will be reduced in the process. As in the case of a vacuum tube amplifier with a single-tuned circuit in its output, the gain-bandwidth product tends to remain constant. Broad-banding the multi-cavity klystron by stagger-tuning the intermediate cavities (i.e., tuning them slightly above and slightly below the input cavity frequency) and loading the cavities to

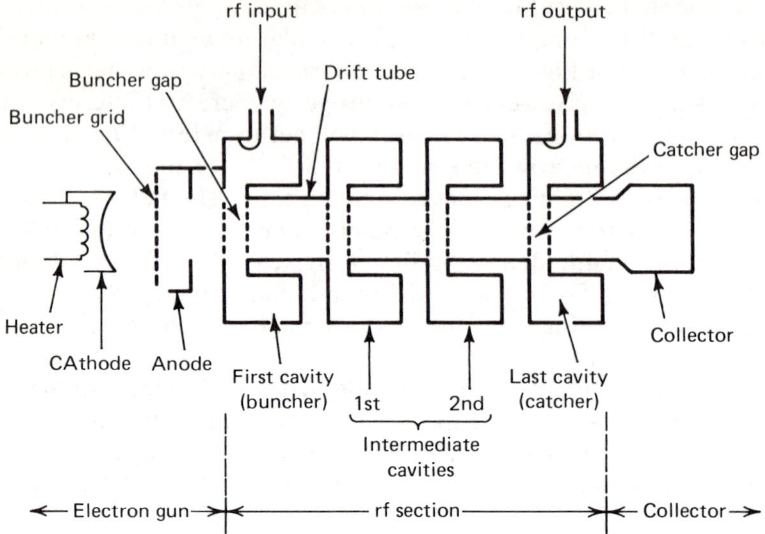

Figure 9-9: Multicavity Power Klystron Amplifier

broaden their response will produce a broadband klystron amplifier, but with reduced power gain. To achieve very wide bandwidths, the slow-wave structure of Sect. 9–2.2 must be used. This would require a TWT design, however. The TWT has an inherently large bandwidth, but it cannot produce the high-peak power outputs that the multicavity klystron has achieved. Pulse klystron tubes have been designed that produce outputs of 30 megawatts at efficiencies of over 40% at frequencies near 3000 MHz.

9–5 THE REFLEX KLYSTRON

If the microwave tube circuit (Klystron amplifier circuit) of Fig. 9-2 is modified by connecting the rf output to the rf input, the tube can become an oscillator. If the net phase shift in the path from the output to the input is adjusted to $2\pi n$ radians (n an integer) and the total gain is adjusted to unity for the loop including the tube gain, the tube will oscillate. Thus, in principle, any klystron amplifier with a positive gain can be converted into an oscillator. However, the two-cavity klystron oscillator is not practical, because when the frequency is varied, the two cavities and feedback loop must be adjusted to restore oscillation conditions at the new frequency. To overcome these limitations, the reflex klystron was developed. This device requires only a single tunable cavity. The proper feedback phase is maintained electronically. Many of the principles of the two-cavity klystron amplifier apply to the reflex klystron.

Figure 9-10 illustrates the principles of operation of the reflex klystron. Electrons emitted by the cathode are accelerated by the accelerating anode and formed into a beam. If oscillation is in progress and an rf voltage V_1 appears across the grids G_1 and G_2, then some of the electrons flowing by will be accelerated, some will be unchanged in velocity, and some will be decelerated, depending upon the polarity of V_1 at the time the electron enters the grid area. The velocity of the stream is thus modulated, as in the two-cavity case. (See Fig. 9-3.) To illustrate this action, consider Fig. 9-11.

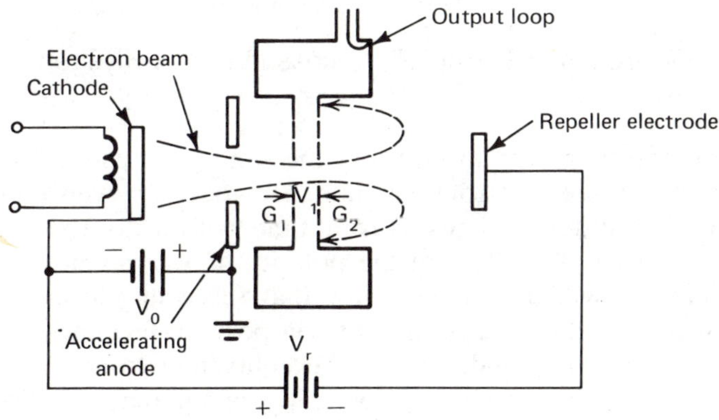

Figure 9-10: Reflex Klystron Principles

Figure 9-11a is a familiar Applegate diagram, but it is different from the previous diagram. (See Fig. 9-4.) Figure 9-11b shows the velocity of an electron leaving the G_1-G_2 area of Fig. 9-10. At time 1 the polarity of V_1 gives maximum electron acceleration. At time 3 V_1 is at 0, and no added velocity Δv is imparted to the electron entering the grid area. At time 5 maximum deceleration occurs, and Δv is a negative value for the electron entering then. From the Applegate time-distance plot of Fig. 9-11a, we see that electron 3 travels through G_1-G_2 without velocity change and is repelled by V_r, causing it to return to G_1-G_2 for collection. This electron is called the center electron. Now electron 1 enters at a time of acceleration, so it leaves G_1-G_2 at a higher velocity than electron 3. It thus penetrates deeper into the repeller field caused by V_r but returns to G_1-G_2 at about the same time as electron 3. Similarly, electron 5 is decelerated and thus leaves G_1-G_2 with a lower velocity. Its trace on the velocity-time diagram of Fig. 9-11b shows little penetration into the re-

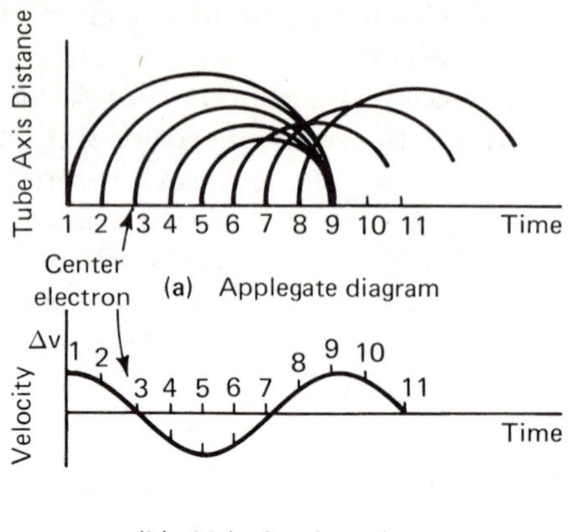

(b)   Velocity-time diagram

Figure 9-11: Bunching Process in Reflex Klystron

peller field, and it returns to G_1-G_2 at the same time as electrons 1 and 3. Clearly, the three electrons are bunched together even though they entered G_1-G_2 at different times. Now at time 9 of Fig. 9-11b the electrons 1 through 5 arrive at G_1-G_2, but the polarity of V_1 provides a retarding electric field. (It will be shown below that this will add energy to the cavity and thus sustain the oscillation via positive feedback.) Referring back to Fig. 9-6, this condition will be equivalent to an electron with velocity $\mathcal{V}_0$ exceeding the phase velocity $\mathcal{V}_{ph}$ of a potential well and being trapped by the well. The electric field across G_1-G_2 of Fig. 9-10 will act like the potential well and absorb energy from the electron. The electrons will be retarded by this field but will thus supply their lost kinetic energy to the field. In summary, in the reflex klystron the electron stream is bunched via velocity modulation, but this bunched stream is returned to the bunching grids in the form of a retarding phase, the latter condition providing positive feedback to maintain the oscillation of the tube. It is also clear that more than one value of repeller voltage V_r could cause oscillation, since the feedback phase $2\pi n$ radians can be achieved using any multiple of 2π radians.

The reflex klystron characteristics of Fig. 9-12 illustrate the multiple-valued oscillation causing features of the repeller voltage: a continuous increase of repeller voltage causes several oscillation modes to form. Each mode satisfies the $2\pi n$-radian feedback condition. The repeller voltage also supplies a fine-tuning feature, but the cavity resonant frequency determines the center frequency f_0. Mechanically changing the cavity dimensions can vary f_0, but fine tuning is invariably achieved via small changes in V_r.

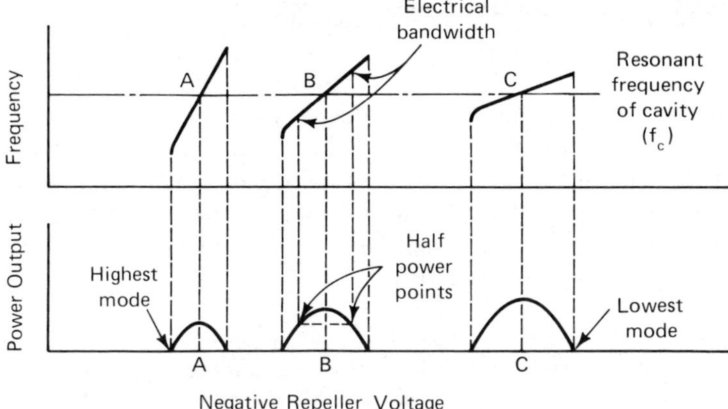

Figure 9-12: Power and Frequency Characteristics of Reflex Klystron

9–6 **THE TWYSTRON TUBE**

The Twystron tube is a hybrid combination of a travelling-wave tube and the klystron tube. The input section of the Twystron is a multicavity klystron, and the output section is a travelling-wave design. The result is a tube with better performance characteristics than that possessed by either component tube separately. With proper design, the Twystron can maintain a constant gain over a wide bandwidth and also have a high-power output. Output powers of 4 to 5 megawatts have been achieved. A schematic diagram representing a Twystron is given in Fig. 9-13.

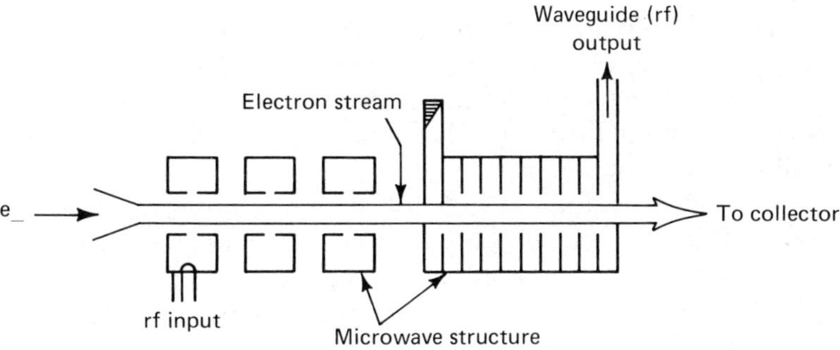

Figure 9-13: Twystron Schematic

9–7 THE BACKWARD-WAVE DEVICE

The last linear-beam (O-type) device to be considered is the backward-wave amplifier and oscillator. Table 9-1 indicates that this is in the slow-wave-structure family of the O-type device. The amplifier configuration will be considered first, as it illustrates the basic concepts involved. Figure 9-14 illustrates, by means of a schematic diagram, the elements of a backward-wave amplifier (BWA).

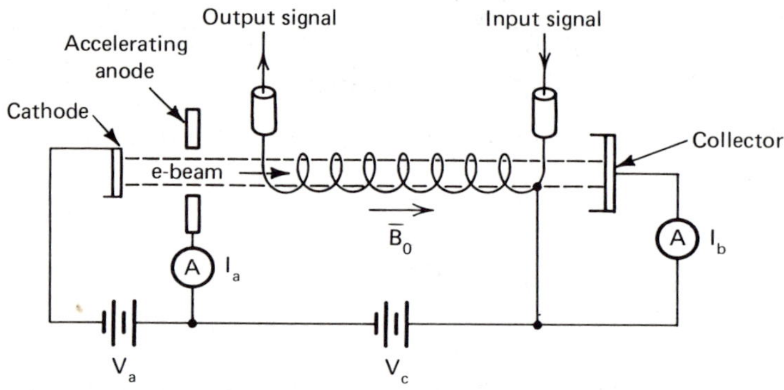

Figure 9-14: Backward-Wave Amplifier Schematic

An electron beam is formed from the accelerated cathode emission by means of the potential V_a. The beam moves parallel to a uniform magnetic field $\bar{B}_0$ that extends the length of the tube; the beam is then collected by the collector electrode. The collector potential V_c determines the beam current I_b, and the potential V_a determines the anode current I_a. This arrangement permits separate control of the beam and helix (collector) currents.

While the input signal travels toward the cathode, the electrons are bunched somewhere near the collector electrode. The bunched electrons interact with the helix slow wave in such a way as to cause the net energy to be transferred from the bunched electrons to the helix field. This permits wave amplification to occur.

In the BWA the phase velocity of the helix wave is opposite in direction to the group velocity of the wave. This mode of propagation in the helix structure makes the backward wave possible. The acual flow of energy, which is in the direction of the group velocity, is from the helix input signal to the output.

This tube should be compared to the M-Carcinotron device (M-BWO) of Fig. 8-16. In the latter device, crossed electric and magnetic fields are used. The interaction with the electron beam is similar, but the

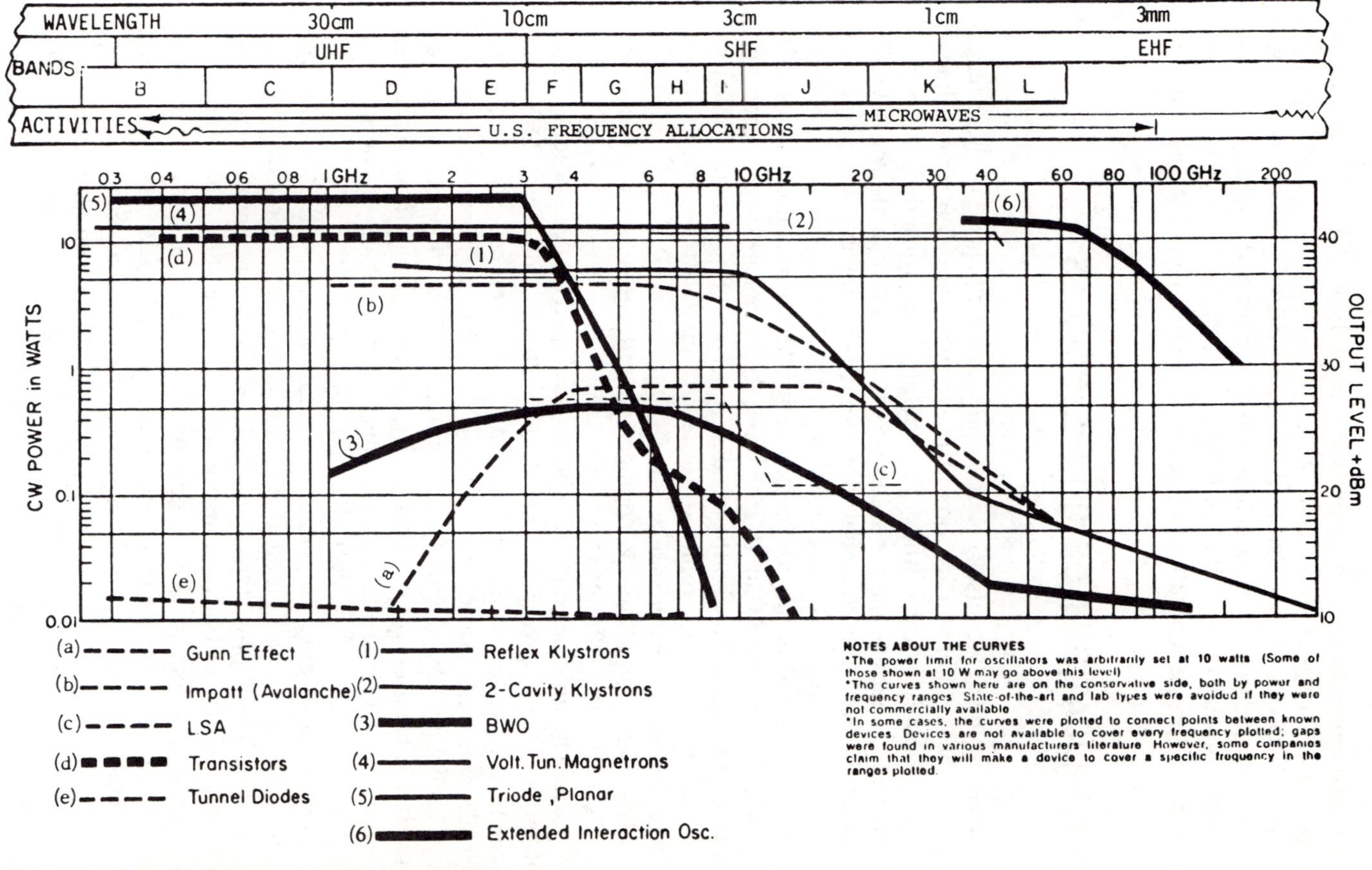

Figure 9-15: CW Power of Microwave Devices *

*Samuel Y. Liao, Microwave Devices and Circuits,©1980, 254. Reprinted by permission of Prentice-Hall, Inc., Englewood Cliffs, NJ.

electric and magnetic fields are opposite that of the O-BWA. No sole plate is required in the O-BWA, of course.

The O-BWA can be converted into an O-BWO by means of a properly adjusted feedback network between the input and output terminals, as in any amplifier device. With a loop gain of unity and a phase shift of 2π n radians, proper conditions exist for sustained electrical oscillations.

9–8 APPLICATIONS OF O-TYPE DEVICES

The O-type device uses electric and magnetic fields, but these fields are generally parallel to each other. These tubes are also termed linear devices, probably because they can operate over wide dynamic signal ranges. They generally operate at lower power levels than M-type devices, but do not achieve the efficiencies of the M-type device. An example of the range of cw power level capability of the O-type device is given in Fig. 9-15. It should be noted that the reflex klystron can cover a broad range of microwave and millimeter wave frequencies. It has been commonly used as a local oscillator in radar receivers. Since local oscillators (LO's) can operate in the milliwatt (or tens of milliwatts) range, the reflex klystron should be useful as a receiver LO up to 50 GHZ. Note also that the solid-state Impatt diode is competitive with the reflex klystron. (See curves (1) and (b) of Fig. 9-15.)

Linear-beam tubes have also been developed that can produce high-peak powers. For example, with a 90 kv supply voltage, 1 MW peak power levels can be obtained from either a klystron or a twystron tube. Typical efficiencies are on the order of 30% for the linear-beam pulse-power amplifier. The usable dynamic range of these tubes can vary from 40 to 80 dB. High-power coupled-cavity travelling-wave tubes have been developed that can produce peak power outputs of up to 200 kw at 2 GHz, but at relatively modest efficiency.

The linear-beam tube tends to generate less noise than the M-type devices and can have a larger bandwidth, along with increased dynamic range. Thus if high power over a large bandwidth is required, a twystron or a TWT design would be suitable. (As we have seen, the slow-wave structure makes this possible, being nonresonant.) The linear-beam tube concept is a superior amplifier, but the crossed-field device is smaller and more efficient, if its other deficiencies can be tolerated. Since the linear-beam device operates at a lower noise level, it is useful as a small, low-power device for receiver applications. For example, low-noise TWT amplifiers are in common use for the sensitive first stages of a radar receiver. The helix-type tube would be preferable for this application, as it has the higher bandwidth.

10   Masers and Lasers

10–1 INTRODUCTION

This chapter is devoted to the description of microwave and milli-meter wave oscillators and amplifiers that are based upon Maser and Laser principles. The acronym Maser stands for Microwave Amplification by Stimulated Emission of Radiation, and the acronym LASER stands for Light Amplification by Stimulated Emission of Radiation. To analyze these types of devices, it is necessary to use nonclassical, or quantum, mechanics. The operation of the O- and M-type tubes of Chapters 8 and 9 were described in terms of free electron charges and the motions of these charges. Chapters 4 and 5 dealt with solid-state electron concepts and the hole and electron motions in a semiconductor. In this chapter a new set of principles applies, the principles of quantum electronics in the various electronic materials. These principles will be described in simple, basic terms, so that the fundamental ideas can be grasped. Since some lasers can operate in the high end of the millimeter band, they must be considered in this text.

Masers operate in microwave cavities, and lasers operate in optical cavities, so that both microwave and optical cavity principles are in-volved. This presents no real problem, as optical waves and microwaves are both electromagnetic (EM) waves and obey the same EM laws. With this in mind, it will be no surprise to learn that masers were developed first and that lasers were derived from them. The laser has had the greater success, and thus, most of the chapter is devoted to laser principles.

10–2 QUANTUM-ELECTRONIC CONCEPTS

The spectrum of EM waves, ranging from AM broadcast waves to the cosmic waves, is given in Fig. 10-1. An EM wave can be described

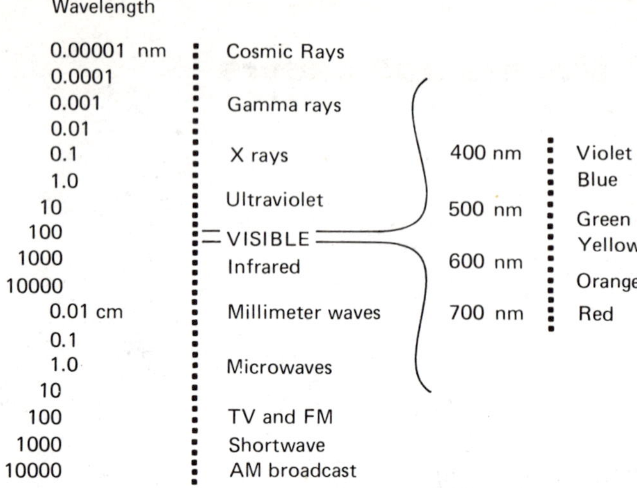

Figure 10-1: Electromagnetic Spectrum

by its wavelength or its frequency. In Fig. 10-1 the wavelengths are given in centimeters (10^{-2}m) and nanometers (10^{-9}m). The range in which optical waves can be seen is the visible spectrum, which ranges roughly from 700 nm to 400 nm. This is a narrow portion of the EM spectrum. Now, an EM wave behaves as both a pure wave and also as a particle. This particulate nature shows up because it has discrete energy and momentum. Figure 10-2 shows that the EM energy exists in wave packets of definite discrete units, or quanta. The wave packet of Fig. 10-2 is called a *photon*. This packet or photon is thought to move through space, thus satisfying the human need to visualize what truly cannot be visualized. The fact that the photon has the properties of a pure EM wave and also the properties of a particle seems contradictory, but it is not; in fact, this duality is basic to quantum mechanics.

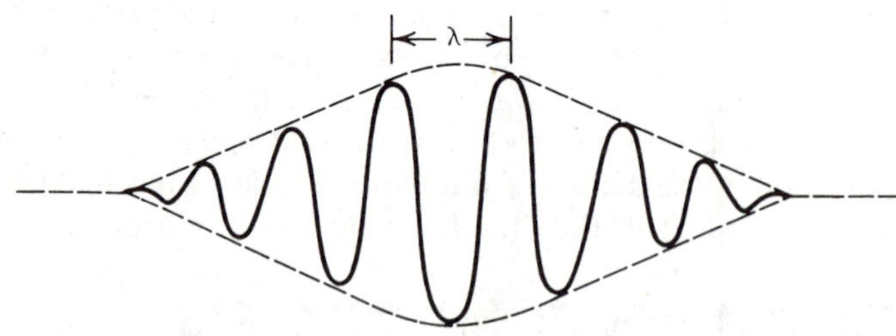

Figure 10-2: Representation of Photon

The energy of a quantum of EM radiation is given by

$$E = h\nu,$$

where

$$E = \text{energy}$$
$$h = \text{Planck's constant}$$
$$\nu = \text{photon frequency.}$$

This formula states that photon energy is directly proportional to frequency, and thus the very high-frequency gamma-ray photons have much more energy then ultraviolet photons. (See Fig. 10-1.) The energy E is often a fixed numerical value for a given material, so that the photon frequency ν is also a fixed numerical value.

Three basic processes occur in a maser material, and these are described in Fig. 10-3. The first process is one in which the incoming photon of energy $h\nu_{12}$ strikes a material that has an electron at the ground energy level E_1. If the next energy state (level) in the material is E_2 and the difference $(E_2 - E_1)$ is equal to the incoming photon's energy $h\nu_{12}$, then there can be an energy interchange in which the photon is absorbed by the material and the electron e_- in the material can move to the level E_2.

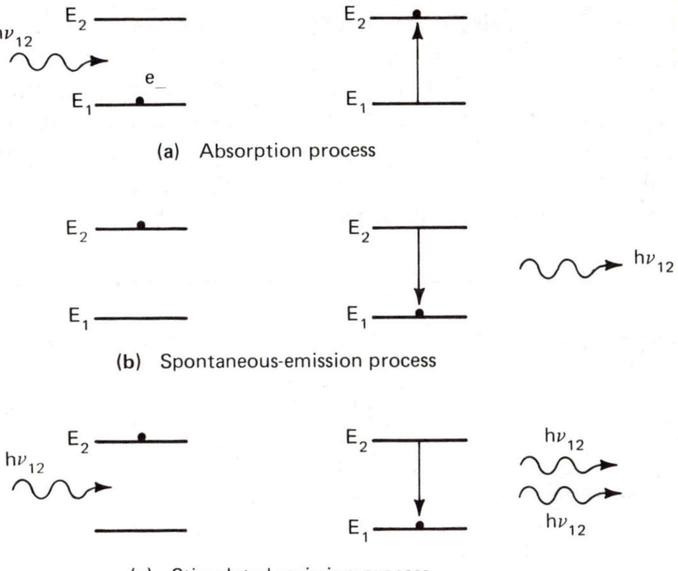

Figure 10-3: Three Basic Maser Processes

When this occurs, the photon disappears, being absorbed by the maser "medium." After a short period of time (10^{-9} to 10^{-3} s), and without an outside stimulus, the electron e_- will spontaneously fall from E_2 down to its normal (ground level) value E_1, with the resulting emission of the photon with energy $h\nu_{12}$. If, instead, a photon of energy E_2-E_1 strikes the medium while the electron is still in state E_2, the stimulated emission occurs. The stimulated emission process is shown in Fig. 10-1c. The incoming photon $h\nu_{12}$ produces a second photon by stimulation of emission, but the new photon is exactly in phase with the incoming photon. The total radiation will be at the same frequency, and all photon emissions will be in phase. Each photon will have the same energy and frequency. Thus, the wave that emerges will be highly coherent, because of this reinforcing action. Einstein's discovery of this stimulated emission process is the basis for all lasers and masers.

In order to produce stimulated emission, it is necessary to have what is called *population inversion* in the laser medium. There must be a significant number of atoms (or electrons) that are raised to the excited state, so that the probability of stimulated emission occurring is correspondingly high. In the normal state in matter the electrons are near their ground state, or lowest energy levels, leaving the upper levels somewhat depopulated. When there are more electrons in the upper population levels than in the ground state, then the population is said to be inverted. These two conditions are illustrated in Fig. 10-4.

The process of raising the electrons or atoms of a material into the inverted population state and maintaining them in that condition is called *pumping*. For a continuous wave (CW) laser to operate, energy must continually flow from the pump into the medium to maintain the population inversion condition. (The same principle applies also to masers, of course.) Pumping may be achieved by many different methods. The pumping source itself need not be coherent (single frequency). Examples of

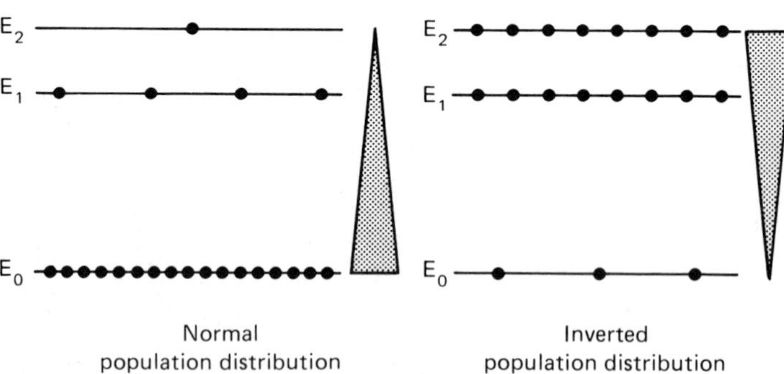

Figure 10-4: Illustration of Population Inversion

incoherent pumping include use of a Xenon flashtube or an electrical discharge. In summary, many techniques exist for pumping, but it can be said that any method successful in causing population inversion in the energy states of the material (assuming the right type of inversion) can permit laser or maser action to occur. Examples of methods of achieving this energy state population inversion will be considered in detail later in this chapter.

10–3 PROCESSES IN THE MASER

10–3.1 The Components of a Maser

An illustration of the components essential to maser operation is given in Fig. 10-5. The illustration is the case of a ruby laser driven by a quartz-tube flash lamp. When the flash lamp is triggered, the photon energy produced raises the atoms in the ruby rod to an excited energy state, producing the required population inversion in the ruby medium. An initial spontaneous decay within the ruby rod excites the optical cavity formed by the mirrors (reflectors) at either end of the rod, and stimulated emission occurs due to a buildup of coherent light energy within the rod-cavity combination. The partial reflector at one end of the rod permits the light energy to escape from the rod. The spacing between the reflectors is chosen to match the frequency ν_{12}, so that a resonant buildup of energy can occur. In other words, the frequency of the stimulated emission from the ruby must match the natural resonant frequency of the optical cavity formed by the reflecting ends of the rod.

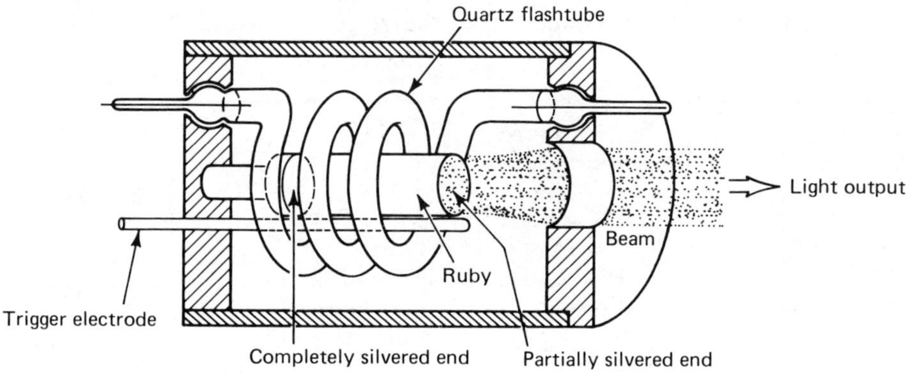

Figure 10-5: Ruby Laser System

The basic components of a laser or maser can then be identified as follows:

1) laser medium

2) pumping source

3) resonant system

4) output coupling device.

10–3.2 Two-, Three-, and Four-Level Laser Schemes

It would appear that a population inversion might be achieved by means of interaction of an rf field of proper frequency with a material having the energy levels of Fig. 10-3. At thermal equilibrium, level one is more populated than level two. An incoming rf wave can increase the population in the second level. It can be shown, however, that when the population in level two becomes equal to that in level one, the stimulated process and absorption process become equal and compensate each other. The material then becomes transparent. This situation is called *two-level saturation*. For this reason, it is impossible to produce population inversion by means of a two-level maser scheme.

The common types of lasers use either three-level or four-level decay schemes. In Fig. 10-6a, the three-level scheme is portrayed. The pump operates to bring about a population inversion of the atoms, raising atoms from ground level (level 1) to the upper level (level 3). Assume that the material is such that an atom in level 3 falls rapidly into level 2, in a fast-decay sense. This action, if continued, can bring about a population inversion between levels 1 and 2. In this three-level scheme, the lasing medium does not become transparent. In the four-level scheme of Fig. 10-6b, the atoms are pumped from the zero level to level 3. The decay

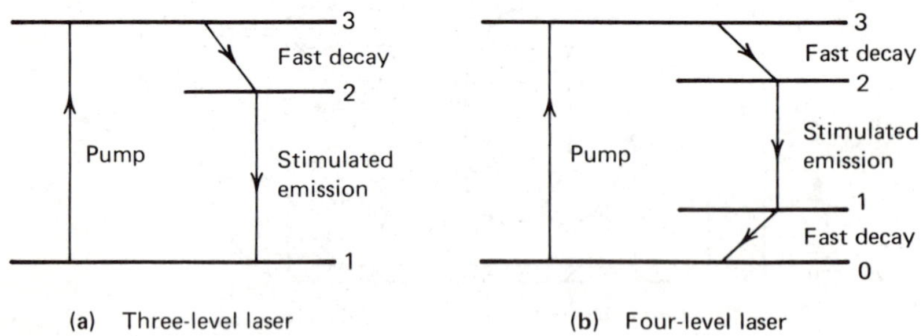

Figure 10-6: Three-Level and Four-Level Laser Schemes

is rapid between levels 3 and 2. For continuous (CW) operation, the transition 1→0 should also be very fast.

It will now be shown that the four-level scheme is more efficient than the three-level scheme. Consider now the three-level case. Since the energy differences in Fig. 10-6 tend to be larger than the thermal levels, nearly all atoms are initially in the ground level.* Begin raising atoms from level 1 to level 3. These will then decay rapidly into level 2. If this decay is fast enough, level 3 will be more or less empty. It is now necessary to raise half of the total population in the material to level 2 in order to have a true population inversion. At this point, any other atom that is raised will then give a population inversion. In the case of a four-level laser, level 1 is also initially empty, and *any* atom that is initially raised to level 3 is *immediately available* for population inversion. It should be clear, then, that a four-level scheme is easier to achieve than a three-level scheme. For this reason, the four-level scheme is preferred to the three-level one in a laser or a maser.

10–4 EXAMPLES OF LASER TYPES

10–4.1 Crystalline Solid-State Lasers

This class of lasers uses an ionic crystal as the active (lasing) medium. The ion in the crystal belongs to one of the series of transition elements in the periodic table, such as Cr^{3+} or one of the rare-earth ions. The ruby laser of Fig. 10-5 is an example in which the crystal of corundum (Al_2O_3) has some of the aluminum ions replaced by a chromium ion (Cr^{3+}). Crystal growth in a molten mixture of corundum and chromium oxide produces this displaced ion crystal.

Another example of a crystalline solid-state laser is one formed of a YAG† crystal doped with the neodymium ion (Nd^{3+}). Sometimes glass doped with Nd^{3+} is also used. The YAG is generally used for continuous (CW) operation, whereas the glass is used as a medium for pulsed lasers. The poor thermal conductivity of glass does not permit useful CW operation.

---

*If the difference between levels 1 and 3 was less than the thermal level kT, then normal thermal processes in the medium would occasionally permit atoms to be raised to level 3 without a pump. If this were the case, then discrete-level separation would not be possible.

†YAG is an acronym for yttrium aluminum garnet

10–4.2 Gas Lasers

A gas laser usually contains a gas at a pressure of the order of 0.3 Torr* that is ionized by means of an ac or dc current. (The pumping is thus achieved via the effects of the electrical current in the ionized gas.) The current travelling through the gas produces free electrons and ions. These charges are accelerated by the electric field and acquire kinetic energy. The free electrons collide with the gas atoms and excite the atoms. The excited atoms can then decay to lower energy states by one of four different processes:

1) collision between electron and an excited atom

2) collision between atoms

3) collision between atom and container walls

4) spontaneous emission.

It is apparent that the production of a population inversion in a gas laser is a more complicated process than that in a solid-state laser, mainly because of the many processes that come into play. It can be said in general that the population will build up in those levels that have the slowest rate of decay into the lower levels. At various current levels and gas pressures, an equilibrium distribution will be reached among these various processes.

Figure 10-7 is a schematic of a gas laser. The arrangement will apply to most gas lasers and is typical for the case of a CW laser. An electrical potential is applied across the terminals of the laser to initiate and maintain the discharge. The two windows at the end of the tube are inclined at the

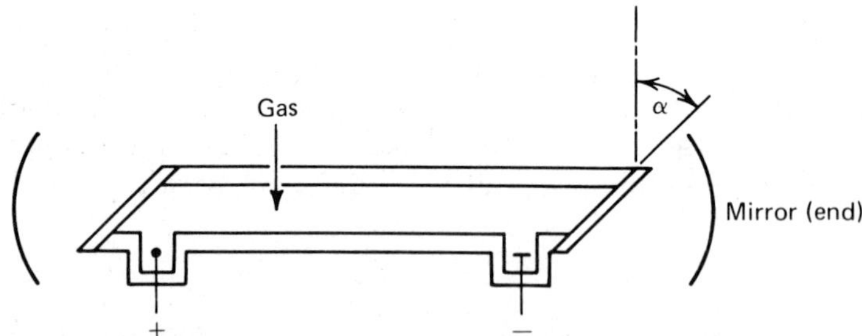

Figure 10-7: Schematic Diagram of a Gas Laser

---

*Equal to a pressure of 0.3 mm Hg

angle α (Brewster's angle). At this angle a laser beam polarized in the plane of the page suffers no reflection losses at the windows. A resonator is then formed between the end mirrors, permitting the laser oscillations to build up at the resonant frequency. In practical cases spherical mirrors rather than plane mirrors are used at either end, since they form more stable resonators.

There are many forms of gas lasers, but they generally fall into four categories: neutral atom gas lasers; ion gas lasers; molecular lasers; and pulsed lasers. These four types of gas lasers are discussed in detail by Svelto*, and thus are not repeated here.

10–4.3 Liquid, or Dye, Lasers

The dye laser has attracted attention because of its flexibility in tuning. The active laser medium consists of solutions of certain organic dye compounds in liquids such as ethyl alchohol, methyl alchohol, or even water. For a given dye, it may be possible to tune the wavelength over a range of about 30 nm, and by changing dyes it is possible to cover any wavelength from the near infrared to the near ultraviolet. These lasers also have a high gain compared to the solid-state laser. Another advantage to the dye laser is the ease with which the active laser medium can be prepared.

The energy-level diagram of a typical dye laser is quite complex, involving many types of energy levels (vibrational, rotational, electronic, etc.). The diagram involves energy levels so close to each other that they appear to be a continuous band. For this reason, the dye laser's tunability is continuous. The dye laser energy level scheme is much more complicated than that of the four-level scheme of Fig. 10-6.

In Fig. 10-8 an incoming electromagnetic wave (photon) can be absorbed by the dye, raising molecules from level S_0 to S_1, as indicated by transition 1. The decay to the lowest level of S_1 is indicated by transition 2, which is nonradiative. Emission can then occur by the permitted transition from the lowest level of S_1 down to some vibrational level of S_0. Nonradiative decay then permits transition down to the ground level of S_0. The process is complicated by transitions from S_1 to the triplet state T_1, as indicated by transitions 3 and 4. This process is referred to as *intersystem crossing*. Some radiation may occur due to some of these processes, but most of the decay from T_1 and S_0 is by collisional losses. The point to be understood in this is that the complicated excitation and decay processes in the dye laser require careful control, since they are

---

*O. Svelto, "Principles of Lasers," Plenum Press, N.Y., 1976. (Chapter 6)

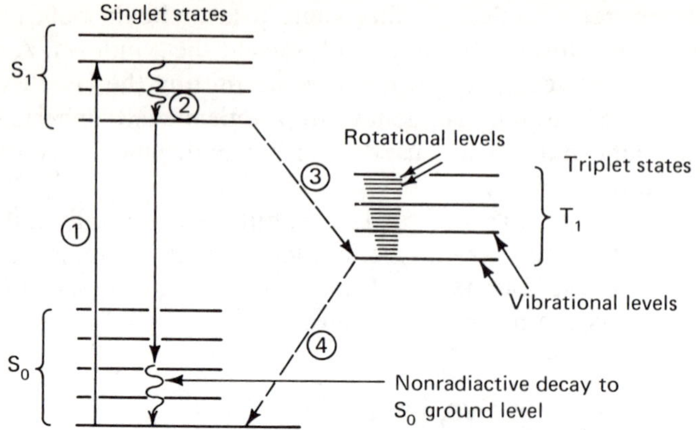

Figure 10-8: Representative Energy Levels for a Dye Laser

more complicated than a four-level laser process. In some cases only a very short pulse operation is possible, due to the rapid decay of the nonradiative transitions.

10–4.4 Chemical Lasers

A chemical laser is defined as one in which the population inversion is a result of a chemical reaction. The chemical reaction is usually a reaction between gaseous elements. The energy released in the reaction is mostly in the vibrational form in the molecules. The laser transitions involved are thus of the vibrational-rotational type. There are two advantages to this laser scheme. The first is that it provides a direct conversion from chemical energy into electromagnetic energy. The second is that high-power outputs are possible, because large amounts of energy are available in chemical reactions.

Two popular types of chemical lasers will be discussed here. They are the H_2–F_2 mixture and the DF-CO_2 mixture. In the first type a chemical reaction between fluoride and hydrogen produces hydrogen fluoride in an excited state, which serves as the laser medium. Laser action takes place between several rotational lines of certain vibrational transitions.

For the case of the H_2, F_2 mixture, the first (or cold) chemical reaction used is

$$F + H_2 \rightarrow HF^* + H.$$

In the second (or hot) reaction the flourine atom is restored by the reaction

$$H + F_2 \rightarrow HF^* + F.$$

The vibrational levels of the HF molecule that become excited are as shown in Figure 10–9, along with the relative populations n(v) that are produced by these two reactions. Laser action takes place between the rotational lines of the vibrational transitions $1 \rightarrow 0$, $2 \rightarrow 1$, and 3–2 (λ = 2.7–3.2μm), as indicated by the arrows in Figure 10–9.

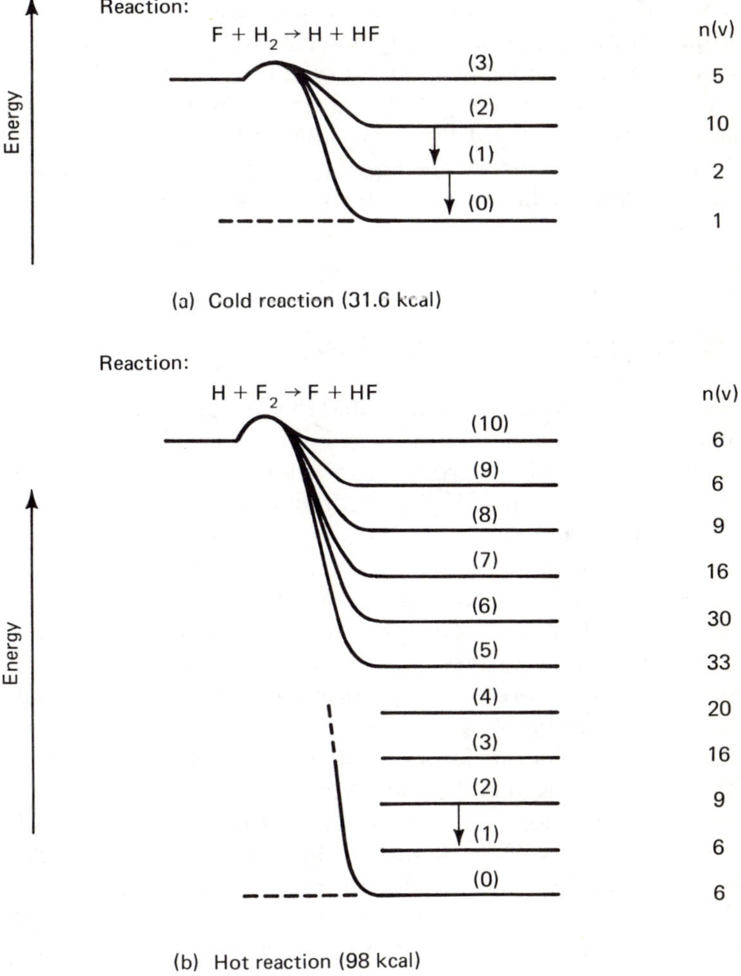

Figure 10-9: Pumping of the Vibrational Levels of the HF Molecule in a Chemical Laser a) Cold reaction b) Hot reaction

In the case of the DF-CO_2 mixture, a hybrid laser action occurs. The population inversion is not produced directly by the chemical reaction. For those interested in the chemical equations, the following reactions are required:

$$1)\ F\ \ \ + D_2\ \ \rightarrow DF^* + D$$

$$2)\ D\ \ \ + F_2\ \ \rightarrow DF^* + F$$

$$3)\ DF^* + CO_2 \rightarrow DF\ \ + CO_2^*.$$

Excited-state deuterium fluoride is produced by reactions 1) and 2). These molecules then excite a CO_2 molecule by resonant-energy transfer, and the excited CO_2 molecule serves as the laser medium. Thus, the laser medium, with its required population inversion, is produced by an indirect chemical reaction, giving it the name 'hybrid laser.'

The above two lasers are popular but have the disadvantage that corrosive products are generated. F_2 is the most corrosive reactive element known to man. In addition, it is hard to dispose of hydrogen fluoride in any quantity.

10–4.5 The Semiconductor Laser

Townes and his students invented the Maser in 1954, but since that time many types of lasers have been built. Over the interval 1957-1961 three papers were written proposing a solid-state laser using a semiconductor p-n junction.* In 1962 a direct bandgap semiconductor laser was demonstrated by Dumke.

The junction semiconductor laser usually uses the III-V compounds from the periodic table. Thus, gallium arsenide (GaAs), gallium antinomide (GaSb), indium phosphide (InP), and other such materials are used.

The semiconductor laser has extreme monochromaticity and high directionality. These are the same properties as that of the solid state ruby laser and the He–Ne gas laser.

The semiconductor laser is quite small, typically being about $50 \times 25 \times 50$ cubic microns in volume (a micron is 10^{-6} m or 10^{-3} mm). The quantum transition ($E_2 - E_1 = h\nu_{12}$) of this device occurs between the conduction and valence bands of the semi–conductor. (See Section 2–2 for details of the energy bands of a semiconductor material.)

Refering to Fig. 10-10a, the gap energy E_g is the approximate quantum energy gap generating a photon of energy $h\nu_g$. In this figure, part (a)

Laser Technology: a survey, National Aeronautics and Space Administration, NASA SP-5116, 1973, by Dr. Paul Zilozer, pp. 2-3.

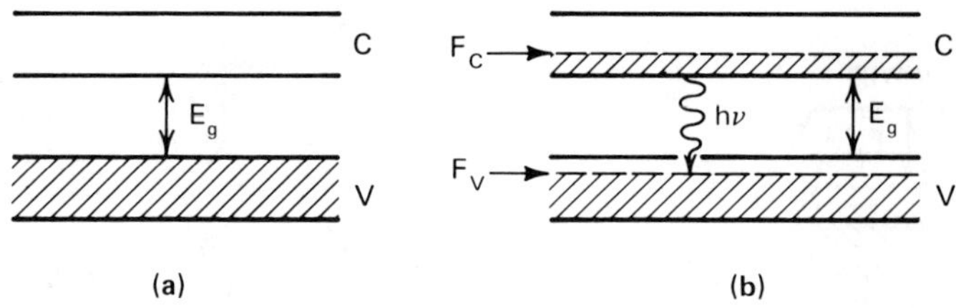

Figure 10-10: Idealized Semiconductor Laser Model

represents the energy states at temperature $T = 0°K$. The valence bands are filled with electrons. Imagine that some electrons are raised to the conduction band. After a very short time (10^{-13} s) the electrons in the conduction band will all fall back into the lowest levels of that band, and all the electrons in the valence band will likewise fall into their lowest unoccupied levels. The valence band will thus have "holes," as indicated in Fig. 10-10b.

Figure 10-10b represents a population inversion in the semiconductor material. The electrons in C (conduction band) will tend to fall back to V (valence band) and the photon will be emitted. If the material is placed in a suitable resonator, laser action can be stimulated.

From an analysis using the statistics of the semiconductor energy levels (Fermi-Dirac statistics) it can be shown that

$$h\nu_1 < F_c - F_v$$

where F_c and F_v are the respective Fermi levels of the conduction and valence bands, as suggested by Fig. 10-10b. It can also be shown that this relationship follows for any temperature where the quasi-Fermi levels F_c and F_v can still be defined.

A very simple way to achieve population inversion is to form a p–n junction diode with heavily doped p– and n–type regions ($\sim 10^{18}$ donor or acceptor atoms/cm^3). Figure 10-11a shows the heavily doped junction case. The Fermi levels in the n- and p–type materials (f_n and F_p) align in equilibrium, as in the abrupt junction case of Fig. 2-3.

With no bias applied, the energy levels across the junction will be as in Fig. 10-11a. When a forward bias voltage is applied, the defusion potential (see section 2–2.3) is nearly cancelled and the energy levels brought to close alignment, as indicated in Fig. 10-11b. The value of ΔF is near to the band gap E_g ($\Delta F \simeq E_g$).

A population inversion is now produced in the "depletion layer" of the p–n junction, as indicated in Fig. 10-11b. The forward bias permits injection into the depletion layer of electrons from the conduction band

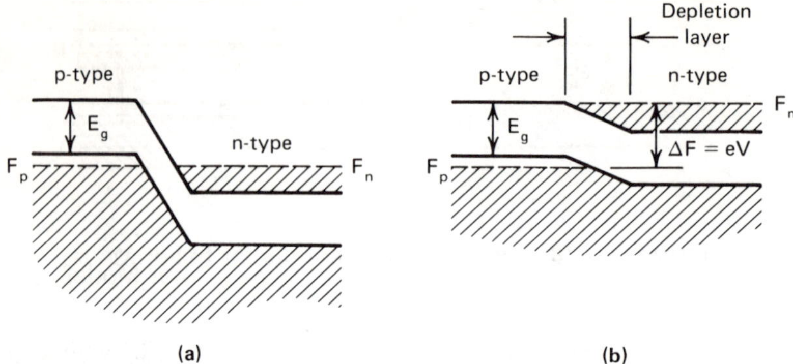

Figure 10-11: Semiconductor p–n Junction Energy Levels

and holes from the valence band, producing the desired population inversion.

A schematic semiconductor laser is given in Fig. 10-12. The shaded region is the depletion layer formed by the potential applied across the junction teminals. Polished surfaces at either end form the required laser resonator. (The rough surfaces diffuse the undesired modes.) The emitted light will be along the direction of the depletion layer, perpendicular to the polished surfaces. The beam out of the device will be a few degrees wide, much wider than the usual laser beam. This will be true because of the small transverse dimensions of the beam.

A widely used semiconductor laser is the GaAs laser ($\lambda = 0.85\,\mu$m). The ratio of emitted photons to the electron–hole injected pairs (the quantum efficiency) is quite high ($\sim 70\%$), giving this laser an overall efficiency of about 30%. The overall efficiency, the ratio of emitted laser power to the electrical power dissipated in the junction, is higher than any other laser.

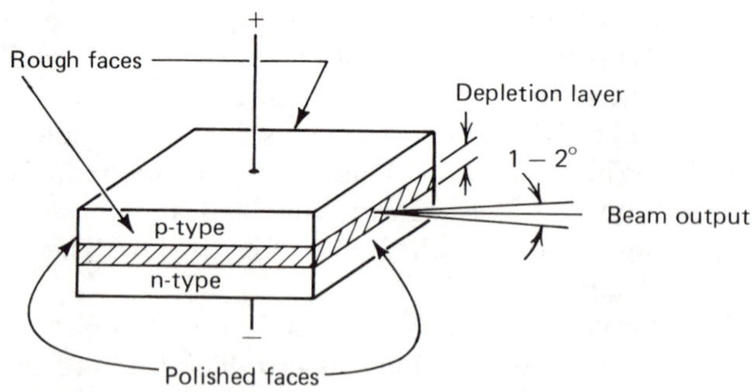

Figure 10-12: Semiconductor Laser Schematic

This form of laser has a very wide oscillation bandwidth ($> 10^{11}$Hz), which makes it quite useful for mode–locked operation. It is also possible to vary the output wavelength continuously by varying the composition of the semiconductor laser material along the axis of the device. The type Ga (As_{1-x} Px) has an output ranging from 0.84 μm to 0.64 μm in wavelength. (The x in the chemical formula represents the position x along the device axis.)

10–5 LASER APPLICATIONS

10–5.1 Present Laser Usage

The laser presently has many highly successful industrial applications. Many problems in production and measurement have been solved by this new knowledge. Laser technology is commercially feasible today, and it is applied in many fields. In the forseeable future many other applications will be found. The laser has found more applications than the maser, mainly because of the high powers that have been obtained.

10–5.2 Examples of Usage

Among the first but still major uses of lasers are the fields of altimetry and rangefinding. Table 10-1 gives a listing of present laser applications, including meteorology, optical communication systems, medical and biological sciences, and direct industrial applications.

Figure 10-13 is a schematic of an optical rangefinder system. This rangefinder was developed as an airborne terrain profilometer. An aircraft

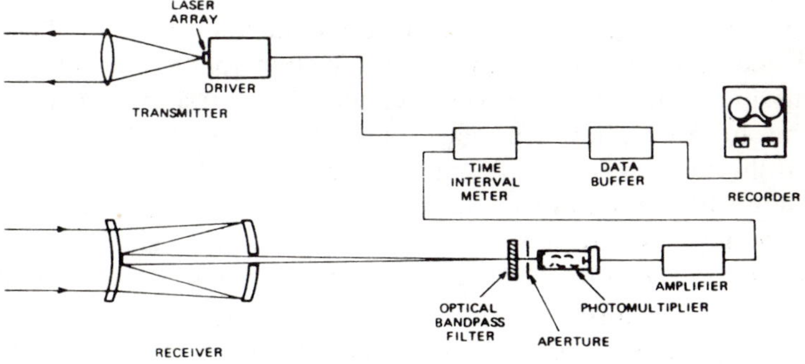

Figure 10-13: Optical Rangefinder System

used it to measure the altitude above ground level in level flight, the profile giving the shape of hills and valleys. The laser recorded vertical range at the rate of 1,000 measurements per second, giving a nearly continuous profile of the terrain when the rangefinder was flown at low speed and at a low altitude.

A rangefinder on the Apollo 15 spacecraft measured the distance from the craft to the moon while in orbit, a distance of about 60 nautical miles. A special reflector left on the moon by Apollo 15 permitted measurement of the distance from the moon to the earth, a distance exceeding 200,000 nautical miles.

Figure 10-14 illustrates the processes involved in the detection of atmospheric pollutants. Infrared (IR) lasers are used to interact with the internuclear vibration–rotation resonances of the atmospheric pollutants. Absorption processes are used in which certain selective laser lines absorb the IR emission. In the case of smokestack effluents, the thermal emission lines are measured by a remote heterodyne radiometer receiver. In other applications, a pulsed IR laser illuminates the pollutant and it flouresces due to the illumination. The flourescence provides a much stronger signal than the usual (Raman) scattering process.

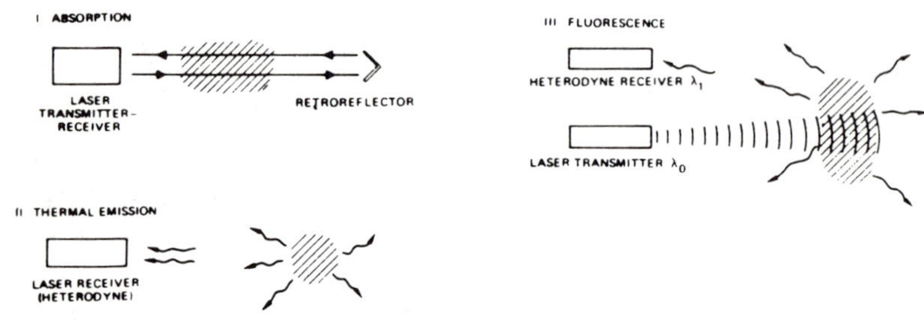

Figure 10-14: Diagrams of Detecting Pollutants Via a Laser Beam

Figure 10-15 illustrates a laser communication system. This is the optical analog of a conventional radio system for voice and television transmissions. These are not yet commericially attractive because of the problems in atmospheric propagation and absorption. They have been successfully tested outside the earth's atmosphere, however.

Figure 10-16 illustrates a Scanning Laser Radar. The Scanning Laser Radar (SLR) was developed by IT & T for the Marshall Space Flight Center. The SLR determined the relative position and orientation between two spacecraft. It also provided basic data to the vehicle guidance and navigation systems, plus pitch, yaw, and line–of–sight angles. The SLR has advantages over conventional microwave radars in that its beamwidth

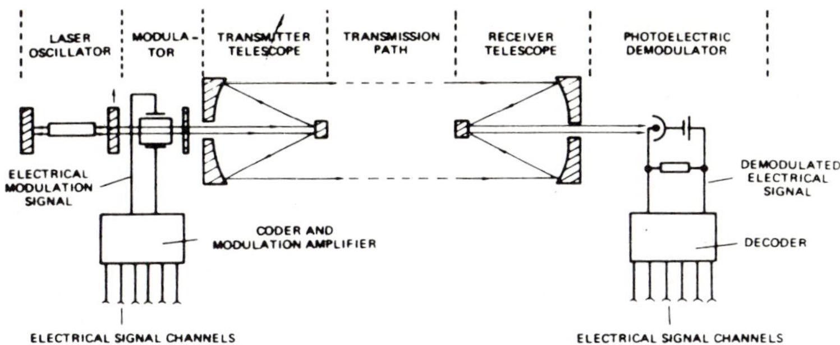

Figure 10-15: Laser Communication System

is much narrower and it has much higher directivity. Because its wavelength is four orders of magnitude smaller than that of microwaves, a much smaller antenna can be used for the same beamwidth.

Of the industrial applications, the technique of laser welding has high interest. Lasers can do welding that is impossible using other techniques. Lasers can repair the inside of a vacuum tube, reattaching filaments and fusing grid leads. Small insulated wires can be attached in electronic assemblies with the advantage that the laser beam vaporizes the insulation without leaving residue. This eliminates the necessity of having to strip insulation from wires before welding.

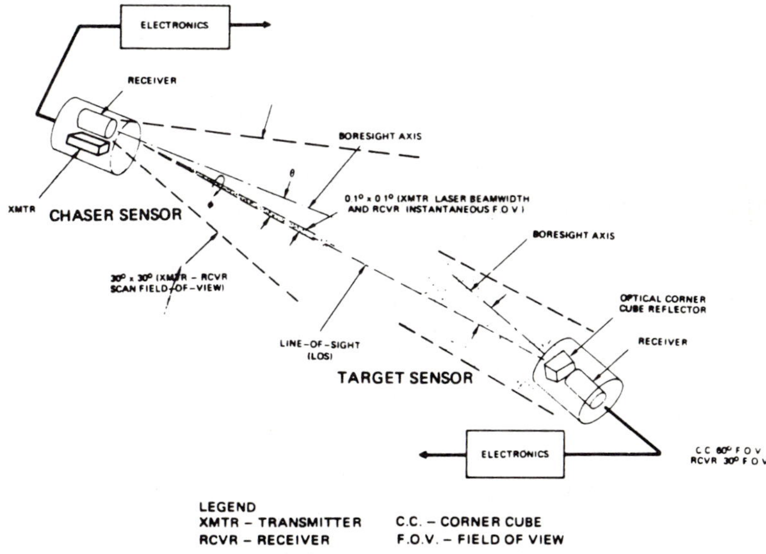

Figure 10-16: Block Diagram of a Scanning Laser Radar

10–5.3 Future Laser Developments and Applications

It can be anticipated that new pumping techniques will be developed with higher efficiency. The old flash lamp technique for pumping a solid state crystal will be replaced by more efficient techniques. Among these will be chemical techniques using the light and heat of certain chemical reactions to induce lasing action. The use of nuclear heat is under study. In this method heat from nuclear fuel is transferred to a cathode and electrons are emitted into a gas lamp or diode which is built to act as a laser itself. This design eliminates the need for an external electrical supply for pumping action.

New laser materials will be developed, some of which will not require external power sources. There are gas lasers today that require no external pumping. The gas dynamic laser uses a supersonic flow of gas to remove waste energy; this is the key to high power devices. The high temperature, high pressure stagnation conditions in the gas flow thermally excite the laser medium. A laser's energy is increased by such sources, and such sources will be of high interest in the quest for nuclear fusion techniques.

Many future laser applications will result in the solution of present technical problems. One possible application is the removal of vortex wakes from aircraft runways. After an airplane has taken off or landed it leaves a vortex wake that can endure for a period of time causing unsafe flying conditions. Crosswinds can extend the life of these vortices. A network of laser beams passing over the runways could detect these vortex wakes, using the deflection of the laser beams to indicate the presence of these wakes.

The Coast Guard is in need of a way to detect the presence of coastal fog banks remotely and automatically. The present videograph backscatter meter becomes unreliable in a fog bank that is not perfectly uniform (i.e., not homogeneous). A laser radar could be used in fog bank detection and in other applications in meteorology.

Future use in space applications are likely considering the lack of atmospheric absorption. A laser beam communication system between satellites would appear to be attractive. The Space Shuttle could well employ laser systems for laser radar ranging and the detection of distant space objects.

Another interesting future application is the development of a color television display using three primary color laser beams. For example, blue and green beams could be produced by an argon (Ar^+) laser and a red beam for a krypton (Kr^+) laser. Very large screens could be illuminated by this means. The TV video would drive (amplitude modulate) the three laser beams and they would be combined and focused onto the screen. A very intense clear image could be produced by this means.

Since the earth to moon distance has been measured to an accuracy of 15 cm via the reflector placed on the moon, there are many very precise distance measurement applications that could be accomplished via laser beams.

Two other applications should be mentioned here. They are holography and optical data processing. Holography was made practical by the laser. Reconstruction of three–dimensional images via holograms has future promise. The technique is promising for the measurement of deformation of objects as a result of variations in temperature and pressure, a property of value in the field of mechanical engineering. In optical data processing, various processes in the filtering of data could be performed more rapidly and economically via the use of laser beams. The process of obtaining Fourier transforms can be done very rapidly, and the cross–correlation between two signals can also be done easily via this method. The contrast in photographic images can also be improved by this technique. Thus future signal processing techniques may be performed by this analog optical technique in preference to the more precise but more expensive digital techniques.

11   Josephson Junctions

11–1 INTRODUCTION

One of the emerging devices not discussed in earlier chapters is a device that will have far-reaching future applications ranging from the lowest electromagnetic (EM) frequencies to the infrared (IR) bands. The device is also one of the most sensitive detectors of EM energy known to man and is accurate enough to be used by the National Bureau of Standards for calibration purposes. The device is sensitive enough to detect the magnetic field generated by the brain. It can also detect the IR signals coming from the galaxies of outer space. What is this device? It is actually a family of devices that is based upon the Josephson effect. The Josephson effect is a superconducting phenomenon that occurs at temperatures of a few kelvins (near absolute zero). Because of the need to operate at such a low temperature, the operation of Josephson devices has been limited to the physics laboratories until the present time. However, the recent development of robust and portable cryostats has made it possible to use Josephson devices in the field. (A cryostat is a low-temperature device (refrigerator) that can cool the Josephson device to below the critical temperature needed to achieve superconductivity in the junction materials.) The ability of the Josephson junction devices to measure voltages, magnetic fields, and magnetic field gradients at or below audio frequencies exceeds any known detector. The Josephson devices have no competitors for sensitivity in these areas. The ability of the Josephson devices to detect frequencies ranging from microwave through submillimeter waves is comparable to the best known detectors, and in the near future may even exceed the other competitive devices.

In the work to follow, the principles upon which these devices are based are presented, along with some projected applications. The practicing electronic engineer should be aware of the potential of these devices, as they are likely to appear in many future electronic systems, some of which will appear in the near future.

11–2 **THE PRINCIPLES OF SUPERCONDUCTIVITY**

In order to understand the principles of the Josephson effect, it is first necessary to review the principles of superconductivity in materials. In a normal conducting material, such as a metal, there are free electrons present that can wander with little resistance through the lattice of the metal's internal structure. These electrons, called *free electrons*, are scattered by impurities in the metal, imperfections in the structure of the metal's crystal formation (lattice defects), and non-EM waves that travel in the metal (phonons). Because of these three scattering sources, the metal has a resistance to the flow of the electrons and the overall electrical current flow at temperatures above the critical temperature T_c. At temperatures below T_c certain metals become superconducting, and the electric current can flow without resistance. (At these temperatures all of the electrons are unaffected by the three scattering effects mentioned above.) Examples of such metals include lead, tin, and niobium.

The phenomenon of superconductivity was first described in a 1957 theory due to Bardeen, Cooper, and Schrieffer (the famous BCS theory). They were awarded the Nobel Prize for this in 1972. In their theory, some of the free electrons in the material formed into what were called *Cooper pairs*, and these Cooper pair electrons were responsible for carrying the resistanceless current (or super-current) that did not result in a voltage drop across the superconductor. (According to Ohm's Law, $V = IR$; i.e., voltage drop (V) is the product of current (I) and the resistance (R) over the path of the current. The natural consequence of zero resistance is a lack of voltage drop in the presence of a current flow.)

The BCS theory was based upon quantum mechanics, a subject that will not be treated here. However, some of the concepts involved are based upon rather simple ideas. Each pair of electrons is said to have a wave function $\psi(\bar{r},t)$ which, according to the symbol used, varies in time (t) and space ($\bar{r}$). The square of this quantity (i.e., $|\psi(\bar{r},t)|^2$) is the probability of finding a Cooper pair at a position $\bar{r}$ in space at a time t. Associated with ψ is the phase of this function. Each electron in the Cooper pair has an equal and opposite momentum, so that the pair has zero momentum. The most important property of these Cooper pair electrons is that they all have the same phase θ, so that they all add coherently, being in exact phase with each other. Because of this, the description of one Cooper pair fits all the Cooper pairs. They thus all add up to a single wave function that applies to all of the minute wave-function pairs in the material.

The above effect is sometimes called a collective effect. The most simple example of this is violins in an orchestra. The intensity of the sound from N detuned violins varies as N. But when all of the violins are exactly tuned and played together, the total sound intensity will vary as N^2. They would then play as a single violin, coherently. The same type of collective effect occurs in the supercurrent phenomenon. The result is due to a lack of randomness, i.e., collective action. When all of the electrons in the material behave in exactly the same way, it becomes a new material with different properties. The new properties include zero resistance and what is called *flux quantization*.

The flux quantization that occurs along with the zero resistance can be easily visualized. Consider a ring made up of superconducting material. Its wave function is given mathematically by

$$\psi\,(\bar{r},t) \,=\, |\,\psi(\bar{r},t)\,|\,\exp\,[i\,\phi(\bar{r},t)].$$

The magnetic flux will be given by the product of the magnetic field in the ring and the ring area. The phase of ψ will vary as one follows it around the ring circumference. One of the required properties of the phase ϕ is that it must change by $2\pi n$ (n being an integer) every time it is traced completely around the ring. This brings about the flux quantization in the ring. The total flux must always equal $n\phi_0$, where ϕ_0 is the fixed-flux quantum. The idea of flux quantization is of great importance in the understanding of the operation of some Josephson effect devices.

Flux quantization can cause a superconducting ring to conduct a constant current. For example, if a conducting ring is cooled in the presence of a magnetic field until it becomes superconducting, the ring will be locked into a state with $n\phi_0$ equal to zero. Now, if the magnetic field is removed, the superconducting ring will remain in the nth state indefinitely, with the creation of a circulating supercurrent that maintains the magnetic flux at the original value. The net result is that the superconducting ring just simply sits there and continues to conduct a current that will flow for years and years. This effect is a consequence of flux quantization in a superconductor.

As another example of the effect of flux quantization in a superconducting ring, consider the case of such a ring that is created in the presence of a zero magnetic field. This sets $n=0$ in the flux quantization, since the total flux must be $n\phi_0$. The ring is thus locked into the $n=0$ state. If a magnetic field is now applied to the ring, it will generate a supercurrent in itself large enough to just cancel the applied field and thus maintain its zero flux state. This supercurrent will persist until the applied field is removed. When the applied field is removed, the supercurrent then returns to zero. In either this case or the above case of a nonzero flux, the ring maintains its flux state (and its n value) once it is made super-conducting.

11–3 BRIAN JOSEPHSON'S DISCOVERY

In a paper published in 1962* Brian Josephson pointed out another consequence of phase coherence in superconducting materials. Figure 11-1 illustrates his argument.

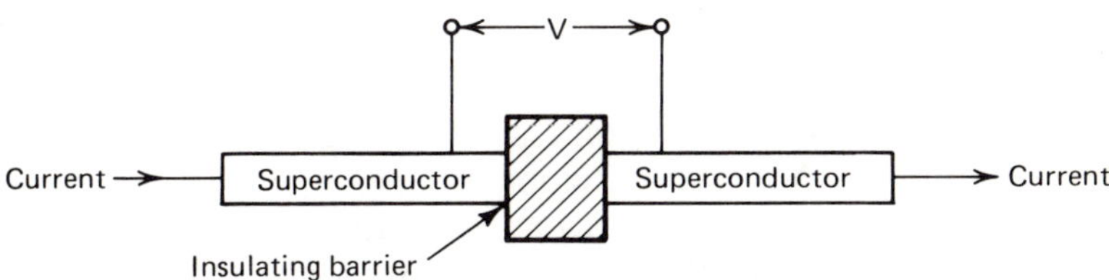

Figure 11-1: The Ideal Josephson Junction

Josephson said that if two superconducting regions are separated by a thin insulating film, then the normal nonzero, finite barrier resistance would be reduced to zero and no voltage V would appear across the junction in spite of the presence of the current flow I, as long as the current density did not exceed some maximum value J_1. Josephson's second prediction was that with finite constant voltages across the junction, the usual dc current would flow, but an ac supercurrent density of amplitude J_1 would also flow and would be at a frequency ν given by

$$\nu = \frac{2\,eV}{h}\ ,$$

---

*B.D. Josephson, "Possible New Effects in Superconductive Tunnelling", Physics Letters, Vol. 1, No. 7, July 1962.

where

$$e = \text{electronic charge}$$
$$V = \text{applied voltage on junction}$$
$$h = \text{Planck's constant.}$$

His theory predicted that a one-microvolt dc voltage would produce a frequency of 483.6 MHz, and that this linear relation between frequency and voltage would be maintained at other dc voltage values. The theory then predicted that a Josephson junction would act as an ac signal source that would have a frequency that would be linearly related to the dc voltage applied across the junction. More simply stated, the frequency would be

$$\nu = 483.6 \, V_0 \, \text{MHz},$$

where

$$V_0 = \text{voltage in microvolts.}$$

The Josephson theory, at its time of publication in 1962, had no experimental verification at the time and was based upon the theory of quantum mechanics. At very low temperatures the junction wave current contributions would all be in phase and would thus add coherently. Thus, the theory predicted a critical temperature above which the material would no longer be superconducting. A critical current density was also predicted above which the phases of the wave current contributions would vary, causing the loss of the super-current effect (i.e., loss of the phasing required for coherent addition of the small current contributions and buildup of the supercurrent).

In his 1962 paper Josephson stated that on the basis of this theory, "effects similar to those considered here should occur and may be relevant to the theory." Josephson's theory* was fully described in his Ph.D. thesis, which was submitted to the University of Cambridge in August, 1964. For this work he shared the Nobel Prize in 1973. His theories proved accurate when tested. Experimental verification is essential to any theory not only because it establishes the validity of the theory but because it provides guidance for an extension of the theory. Josephson himself must have recognized this need for verification, as the 1962 paper was

---

*B.D. Josephson, "Supercurrents through Barriers", Advances in Physics, Vol. 14, p. 419 (1965).

entitled "Possible New Effects in Superconductive Tunnelling." The full impact of this discovery on millimeter wave and computer systems has yet to be seen.

11.4 THE BASIC JOSEPHSON EFFECTS

The 1962 theory of Josephson predicted two major effects and suggested a third. The two major effects are the so-called dc and ac Josephson effects. The third effect is an extension of the ac effect that has received little attention. The two major effects, plus some consideration of the third less-recognized effect, are discussed in the work to follow.

11–4.1 The Dc Josephson Effect

Figure 11-1 illustrated the principle of the dc Josephson effect. Two superconducting materials are separated by a thin insulating barrier, the thickness of which is 10 to 20 angstrom units. Above the critical temperature T_c, the two conductors are normal in their behavior, and the barrier might have several ohms of resistance. Josephson predicted that when the metals were cooled to below T_c, the barrier resistance would vanish and a supercurrent could be passed through the junction without a voltage drop. The cause of this involves what is called *quantum mechanical tunnelling*: the Cooper pairs of the superconducting material are said to tunnel from one superconductor to the other through the barrier, maintaining their phase coherence. Throughout the junction the phase is constant, even without an applied field or current. However, when an external current is passed through the junction, there is a phase change across the barrier, $\delta\phi$. The external current is related to this phase by

$$I = I_e \sin(\delta\phi),$$

where

$$I_e = \text{maximum supercurrent flow.}$$

If the external current exceeds I_e, a voltage will appear across the junction. According to Josephson, this will occur when the screening effects in the tunneling process are cancelled. This cancellation comes about due to the fluctuation of the phases of the supercurrent elements in the barrier, a process that prevents the coherent addition of the Cooper pair wave functions.

Each Cooper pair of free electrons has a wave function described by

$$\psi(\bar{r},t) = |\psi(\bar{r},t)| \exp[i\,\phi(\bar{r},t)].$$

Superconductivity occurs when all of the Cooper pair wave functions have the same phase ϕ. In this case they all add together, and one wave function describes all of the Cooper pairs. The result is a flow of supercurrent, an effect that requires no voltage. At a certain maximum current density level in the junction, the phases of these wave functions will not add up coherently in the barrier, a voltage develops across the junction, and the dc Josephson effect is lost.

11–4.2 The Ac Josephson Effect

When a voltage is placed across the Josephson junction of Fig. 11-1, the supercurrent in the junction will oscillate with time. This phenomenon is the ac Josephson effect. As mentioned previously, the frequency of the oscillation is related in a simple way to the voltage applied across the junction. This relation can be written as

$$\nu = \frac{V}{\phi_0} \text{ (Hz)},$$

where

$$V = \text{applied voltage in volts}$$

$$\phi_0 = \frac{h}{2e}$$

$$e = \text{electronic charge}$$

$$h = \text{Planck's constant.}$$

As far as is known today, this relation is exact. The oscillation persists up to voltages of several millivolts and results in frequencies up to 10^{12} Hz. This relationship is so precise that it has been used by the National Bureau of Standards and a number of other laboratories to maintain the standard of electromotive force for the volt. The fundamental value of the ratio of e/h has been revised, based upon this effect. This is possible because a precise frequency measurement is translated into a precise voltage measurement by the above equation. ϕ_0 is one of the basic measurement quantities of physics. For these reasons, the above equation is called an exact equation; it is a very precise relationship.

11–5 DEVICE LIMITATIONS

11–5.1 Frequency Limits

The frequency range of the ac Josephson effect is very broad. The upper limit is over 10^{12} Hz, or 1,000 GHz. Since this can be accomplished by the application of about 2 millivolts, modest potentials are involved. Frequency sweeps from a fraction of a GHz to 1,000 GHz could be covered by a voltage varying from a microvolt to several millivolts. The precise linear relationship between voltage and oscillating frequency makes this possible.

11–5.2 Switching Times

The Josephson junction, since it can oscillate at extremely high frequencies, can also operate as a very rapid switch. The device has an estimated switching time of 10 ps (10^{-11}s). This is ten times faster than today's best computer element capability. For this reason, the device is being considered for use in a new type of ultrafast, ultrareliable computer.

11–5.3 Other Limitations

A main limitation of Josephson technology today is the need to cool the devices to about 4.2°K, the temperature of liquid helium. The junctions are usually immersed in this liquid. Traditionally, low-temperature operation requires a double-walled Dewar glass flask. The inner Dewar containing liquid helium is surrounded by a second Dewar containing liquid nitrogen. More robust Dewar arrangements have been recently made that are of metal or fiberglass. The fiber Dewar glass has the advantage of being nonmagnetic, an important consideration for magnetic field measurement. A typical cryostat can hold one charge of helium for several days, consuming as little as one liter per day.

Another limitation is the need for very precise uniformity of the tunnel barriers when many devices are used, such as in computer applications. The development of sputter procedures to grow oxide barriers has overcome many of these limitations, since this method allows oxide barrier growths of thicknesses of tens of atomic diameters.

11–6 ELECTRONIC SYSTEM IMPLEMENTATION OF THE JUNCTION

11–6.1 The Precise Nature of the Junction

The precise nature of the Josephson junction has made possible its use as a standard in national laboratories. This precision has been used to maintain the standards of emf. (electromotive force) at the National Bureau of Standards.

As mentioned earlier, the device can be used as a very-broadband-tunable oscillator, since the oscillation is precisely voltage-controlled.

What has not been discussed yet is the device's ability to measure very small voltages, magnetic fields, and magnetic grandients. This brings up such odd-sounding device names as SQUIDS, SLUGS, etc. In some of these applications the Josephson device has no competitor. It is simply the most sensitive detector device known.

11–6.2 Use As a Magnetic Field Detector

The basic detector for sensing low-frequency signals is the super-conducting quantum interference device *(SQUID). This device uses a combination of Josephson tunnelling and the flux quantization effect. There are two types of SQUID, namely, the dc and rf types.

Figure 11-2a illustrates the basic dc SQUID configuration. Two Josephson junctions are mounted on a superconducting ring. The ring requires quantized flux states, as mentioned previously, as the total ring flux must always equal n ϕ_0. (See Sect. 11.2.) The critical current I_c in the ring will vary with the external (applied) flux as shown in Figure 11-2b. The period of the value of I_c is the familiar flux quantum ϕ_0. If an alternating current is applied to the SQUID that periodically exceeds I_c,

---

*The introduction of material on the SQUID is necessary as it is being used in millimeter wave bolometer detectors. Understanding the SQUID is necessary in order to understand the very sensitive SQUID-bolometer of section 11.6.5.

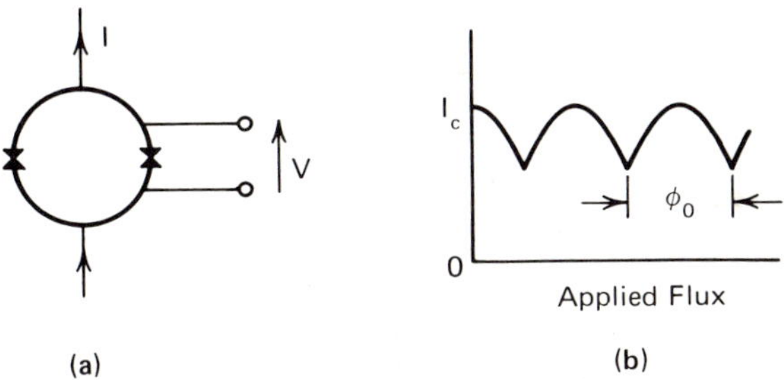

Figure 11-2: The Dc SQUID

then, periodically, a voltage will appear across the weak links, and periodically no voltage will appear across the superconducting-state links. The volt-ampere characteristics of the SQUID change, due to a coupled, applied flux, as indicated in Fig. 11-3.

In the figure, if the flux applied to the SQUID changes by a half quantum ($\phi_0/2$), the dc voltage across the SQUID will vary as shown, producing the voltage change δV. In this case a value of current I_0 is chosen that exceeds the critical value. The amplitude of voltage for this condition is typically a few microvolts, and thus, conventional, room-temperature electronics can be used to amplify the signal. Thus, in practical applications the dc SQUID is operated with a bias current that always exceeds the critical current level I_c.

The rf SQUID uses a different geometry, as indicated by Fig. 11-4. A single Josephson junction is mounted on a superconducting ring. A steadily increasing magnetic field induces steps (quantum transitions) in the SQUID flux. The junction is periodically driven by the current I_g into the voltage-supporting junction state. There will then be a circulating supercurrent in the ring that will be periodic with the applied flux, with period ϕ_0. This will be detected by coupling the SQUID to the resonant LC circuit of Fig. 11-4, the coupling occurring between the ring and the inductance of the LC circuit. Typically, the resonant frequency is 30

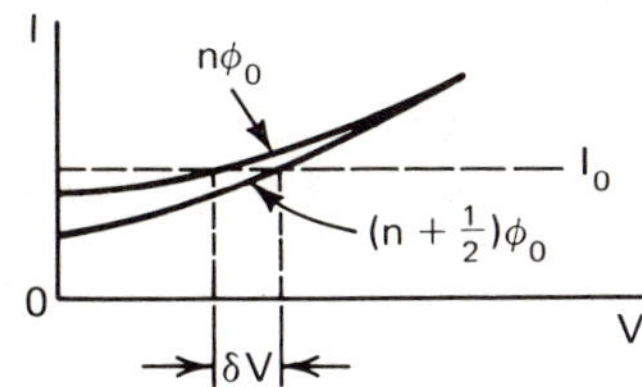

Figure 11-3: SQUID I-V Curves

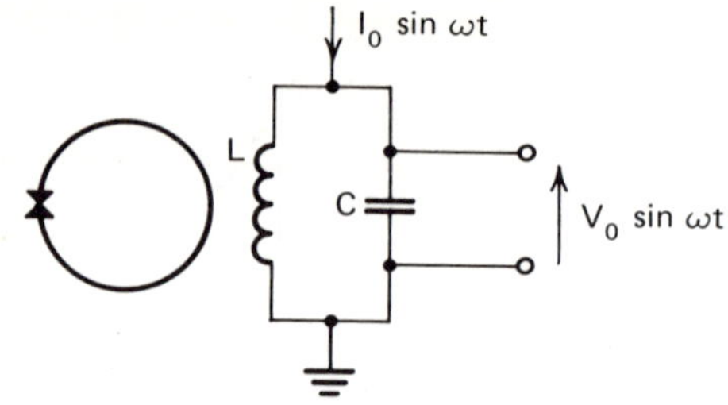

Figure 11-4: The Rf SQUID

MHz. The amplitude of the rf voltage V_0 is proportional to the circulating supercurrent in the SQUID and is therefore periodic in the flux applied to the SQUID. The rf voltage is amplified and rectified by conventional electronics.

Both the dc and rf SQUIDS produce output voltages that are periodic in the applied flux, as in Fig. 11-2b. In practice, though, these signals are put into a feedback circuit that results in a voltage output that is linearly proportional to changes in the applied flux. As an example, Fig. 11-5 shows a feedback circuit used in a dc SQUID.

The bias current I_0 is as described before. The amplifier is a conventional dc amplifier. The signal output is fed through R_f into the SQUID via the coupling coil L_f and the feedback current I_f. The action of the feedback loop is to cancel out the effect in the SQUID of any applied field ΔH, via the feedback current I_f. Since the output voltage of the amplifier is $I_f R_f$, the output is linear with the input signal that produced ΔH. The role of the SQUID is thus one of a null detector in the feedback loop system. The same principle is applied to the rf SQUID, so that the output is now proportional to the input signal flux ΔH.

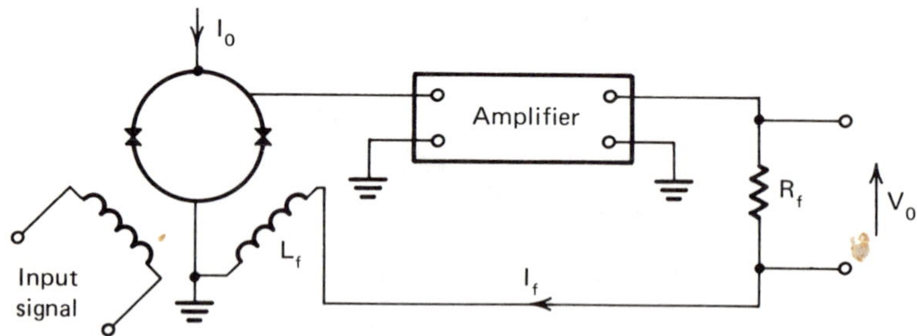

Figure 11-5: Typical Feedback Circuit for a Dc SQUID

11–6.3 Flux Transformers and the Gradiometer

The sensitivity of the Josephson magnetic field detector can be enhanced considerably by the use of a superconducting flux transformer. The principle is illustrated in Fig. 11–6a.

The flux transformer consists exclusively of superconducting elements. (All of its transformer elements are supercooled.) The pickup loop of Fig. 11-6a is the primary winding of the transformer, and the secondary winding is closely coupled to the SQUID. The SQUID and the secondary winding are enclosed in a superconducting can. The magnetic field to be measured is intercepted by the pickup loop. This field generates a persistent current in the loop that drives the secondary loop inside the shield can, inducing a flux in the SQUID. This transformer action can result in very high field sensitivity in the SQUID. Fields as small as 10^{-10} gauss/ $\text{Hertz}^{+1/2}$ have been detected (resolved) by using this technique.

With the very high sensitivity mentioned above, the limiting factor in magnetic field detection becomes the noise in the earth's magnetic field. To overcome this obstacle, it becomes necessary to operate in the gradiometer mode. The gradiometer does not detect the uniform magnetic field, but detects the gradient in the local magnetic field. (The gradient refers to the change in field as a function of position and can be described by $\dfrac{\partial H}{\partial r}$, i.e., the change in magnetic field as a function of position along the direction r.) Figure 11-6b illustrates the gradiometer design for the transformer-SQUID system. The transformer loops must have the same areas for proper gradiometer operation. The loop connections are crossed so that only the difference in the currents is sensed by the SQUID ring. The net supercurrent is then proportional to the difference between the two field values, thus measuring the local field gradient between positions Z_1 and Z_2 in the figure. (Actually, the measurement is a comparison of

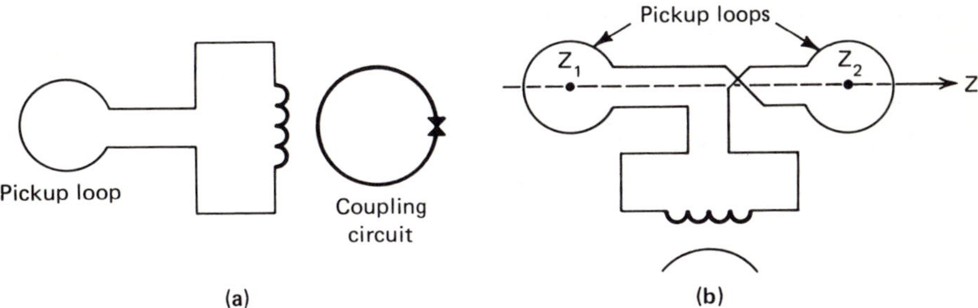

Figure 11-6: Flux Transformer and Gradiometer Circuits

the average field enclosed by the loop encompassing point Z_1 and the loop encompassing point Z_2. Mathematically, this can be expressed as

$$\frac{\bar{H}_{Z_1} - \bar{H}_{Z_2}}{Z_1 - Z_2} \simeq \frac{\partial H}{\partial Z},$$

where the bars indicate mean, or average, values.) The full field gradient requires field measurements to be made in all three coordinate directions, namely $\frac{\partial H}{\partial x}$, $\frac{\partial H}{\partial Y}$, and $\frac{\partial H}{\partial Z}$. Three component gradiometers (or vector gradiometers) have been built and used with success.

The sensitivity of the gradiometer is quite high. The first published account of gradiometers reported a resolution of 10^{-10} gauss/cm/Hz$^{1/2}$. A multihole SQUID and a resonant circuit frequency of 300 MHz were used.

The gradiometer tends to discriminate against signals generated at a distance. Since the main sources of non-man-made magnetic noise are the upper atmosphere and the fluctuation in the earth's field, the gradiometer is relatively free from response to these major magnetic noise sources. But the same principle also applies to distant man-made sources of noise. For this reason, most of the practical SQUID sensors are gradiometer designs.

11–6.4 Use As a Voltmeter/Thermometer

The SQUID can be used to sense very small voltages. Figure 11-7 shows the connection used to detect minute potentials. A resistance is placed in series with the superconducting coil coupled to the SQUID ring. (Connection could also be made directly to the niobium SQUID ring, as in Fig. 11-5, so that the connection a-b of Fig. 11-7 would become the input amplifier terminals of Fig. 11-5.) The signal voltage V_s generates a current I_s that is detected by the superconducting sensor. The voltage

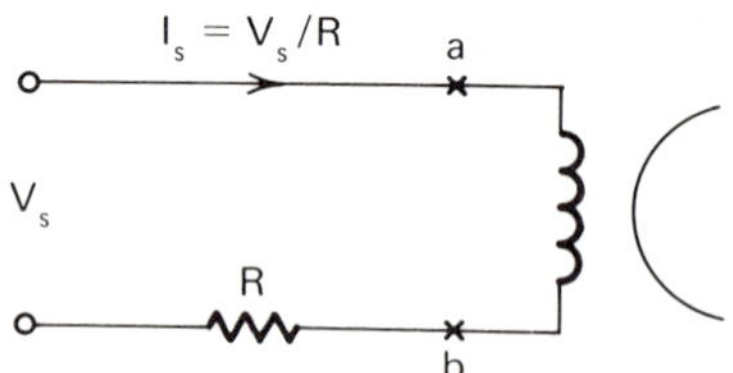

Figure 11-7: The SQUID Voltmeter

sensitivity is quite high for this device. A sensitivity of 10^{-15} volt/Hz$^{1/2}$ is readily achieved. The limitation of this detector is the Johnson noise generated in the resistance R. For high sensitivities, values of R as low as 10^{-8} ohms have been considered.

The above device can also be used as a very sensitive thermometer. The V_s terminals of Fig. 11-7 are shorted out so that the signal is the noise generated by the resistance R. Since the root-mean-square current in R is proportional to the absolute temperature T, the SQUID can now measure temperatures. Temperatures down to a few thousandths of a Kelvin have been measured by this method.

11–6.5 Use As a Millimeter Wave Dectector

Josephson detectors can be used at microwave and millimeter wave frequencies to detect signals in that frequency range. A single microwave frequency coupled to a point contact junction will give the step response I-V characteristic of Fig. 11-8. The steps of constant voltage are given by

$$|V_n| = \frac{nhf}{2e},$$

where

$$n = \text{an integer}$$
$$h = \text{Planck's constant}$$
$$e = \text{electronic charge}$$
$$f = \text{microwave frequency.}$$

The height of the current steps will oscillate with the increase of the microwave input power. The operation of the detector is complex, as the junction will oscillate at a frequency that depends upon the voltage ap-

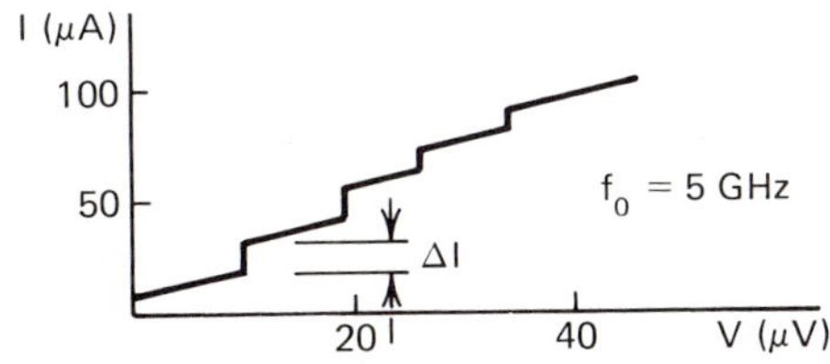

Figure 11-8: I-V Characteristics of a Point Contact Josephson Junction

pearing across it. Wherever $|\nu \pm nf| = 0$, there is a zero frequency beat which results in the current step. When the energy difference between Cooper pairs across the junction (2eV) is equal to nhf, pairs can cross the barrier coherently, with the emission or absorption of n microwave photons. Thus, in the figure, the voltage steps occur when the Cooper pair energy resonates with the quantity nhf. In terms of elementary quantum concepts, the incoming microwave energy consists of many quanta of EM photons of energy hf. The energy difference the Cooper pairs see across the resonant junction is $2eV_n$ and, since the junction is at the resonant or step condition, the Cooper pair energy is some integral multiple of the microwave photon energy hf. The larger step voltages V_n correspond to higher multiples of step n, and the relationship is "quantized."

The microwave energy generates constant current steps in the I-V response of Fig. 11-8. However, the critical current in the junction will vary in an oscillating way with the microwave field. The critical current will correspond to the n = 0 induced step. (Note that for n = 0 in the equation V_n, the condition for Cooper pair resonance is satisfied for all frequencies f.) But the critical current I_c will vary with microwave frequency f. For small signal levels,

$$I_c \, \alpha \, \frac{V_s^{\,2}}{4 \, \pi^2 \, f_s^{\,2}} = \frac{V_s^{\,2}}{\omega_s^{\,2}} \, ,$$

where

$$V_s = \text{voltage induced across the junction by radiation at } f_s$$

$$\omega_s = 2\pi f_s = \text{microwave radian frequency.}$$

The junction is thus a square-law detector, responding to the power in the radiation. Notice that the junction loses sensitivity at higher frequencies.

In practice, the point contact detector is mounted in waveguide, and a constant current I_0 is used to bias the junction at a nonzero voltage. The critical current I_c in the junction will be modified by incoming microwave radiation. The I-V characteristic will also be modified, due to ΔI_0, and the voltage across the junction V_s will be modified, too. The detector will generally respond to a very wide range of frequencies, and thus, some prefiltering of the microwave frequency is needed in order to determine the energy in the prefiltered band. A common technique is to sweep the incoming signal band (for example, by an interferometer) to measure the spectrum of radiation.

The present range of operation of Josephson microwave detectors is in the vicinity of 1 mm (300 GHz) band. The sensitivity of these detectors is comparable to the best semiconductor broadband bolometers.

Josephson detectors have also been used as heterodyne detectors, mixing a local oscillator (f_{LO}) with the signal to be detected (F_s). Usually, the incoming signal is close in frequency to the local oscillator frequency, i.e., $f_{LO} \simeq f_s$. The local oscillator amplitude is much larger than the signal to be detected. Figure 11-9 shows the effect on the I-V characteristic due to the input signal.

The input signal will modulate the I-V characteristic, as indicated by the dashed lines of Fig. 11-9. A bias current L_0 is indicated that is between the n = 0 and the n = 1 step voltage. The bias current exceeds the critical current I_c and thus a voltage is developed across the junction. For the heterodyne case, the I-V characteristic curve will fluctuate as indicated in Fig. 11-9, but this fluctuation will be at the frequency $f_I = | f_{LO} - f_s |$, which is the difference frequency. The difference frequency is then detected by a conventional intermediate frequency amplifier and detector scheme.

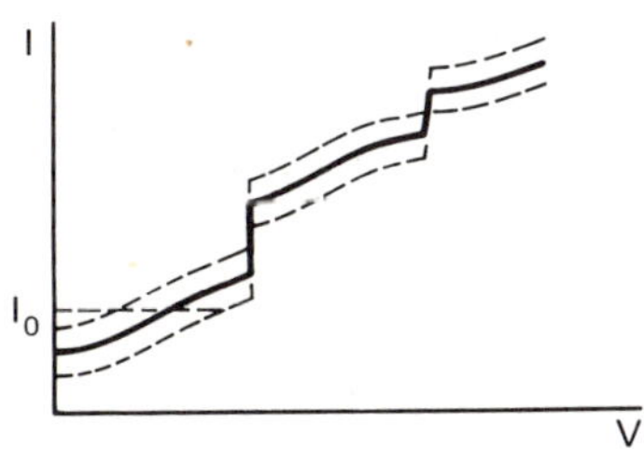

Figure 11-9: Heterodyne Mixing and the I-V Characteristic Curves

11–6.6 The Broadband Bolometer

The bolometer is a device that responds to a temperature change in a material that is induced by the material's absorption of microwave or millimeterwave energy. The bolometer element simply absorbs microwave energy and is heated by this process. The bolometer has some temperature-sensitive parameters that can be used to sense this temperature change. Sometimes the resistance in the bolometer is used as the parameter to be sensed. A Josephson junction has been devised that senses the temperature rise in a bolometer element. A thermal link is used between the Josephson junction and the bolometer element, the link being a low-heat-capacity sapphire substrate. The critical current in the junction increases exponentially as the temperature is lowered. Thus, a small change in the sapphire substrate's temperature produces a large change in the critical current of the junction. A SQUID is then used to sense the change in the critical current of the Josephson junction.

Sensitivities of the order of 10^{-15} watt/Hz$^{1/2}$ have been achieved. The wavelength range of 0.1 to 1.0 millimeters seems to be preferred, making this a good submillimeter wave device. Should this be the case, this would be the most sensitive broadband submillimeter (far-infrared) radiation detector available today.

The bandwidth of this bolometer would be limited only by the efficiency of the coupling between the bolometer element and the millimeter wave signal source. The choice of a substrate material with low heat capacity (such as sapphire) will result in a fast-response device.

11–7 SYSTEM APPLICATIONS

There are many possible applications for the Josephson junction in the future. It appears now in many forms, depending on the particular quantity to be sensed. The junction is a very precise sensor, as emphasized by its use by the National Bureau of Standards. The use of Josephson junctions in computers will offer the possibility of the developement of increased speed and reduced size. The use in SQUIDS will open new vistas in the measurement of low-frequency voltages, magnetic field, and magnetic field gradients.

An important and significant future use of Josephson-type devices is in the generation and detection of millimeter and submillimeter EM signals. Using the ac Josephson effect, the device could generate a very wide range of millimeter wave and submillimeter wave signals, being swept in frequency by simple dc bias voltages. As a detector, it can be used over very broad bandwidths, employing the bolometer detection principle. As a heterodyne detection device, it can be used to translate the incoming signal (perhaps at 300 GHz) down to conventional IF frequencies for simple detection. The device has high promise for wide use in the submillimeter wave bands, due to its high sensitivity and wide bandwidth potential.

12 High Power at Millimeter Wavelengths, or The Gyrotron Tube

12–1 INTRODUCTION

The O-type and M-type devices of Chapter 8 and 9 are the principal microwave oscillator and amplifier tube types. They are limited, however, by the necessity of having resonators and microwave structures that are limited to dimensions comparable to a wavelength. Attempts to scale these designs into the millimeter wave region are not successful, however, for several reasons. For example, if an M-type magnetron is scaled down for use as a 40-MHz oscillator, the cavity dimension must be comparable to a wavelength, namely 7.5 mm. Such small cavity dimensions require the power density in the tube to be very high, limiting the power output to much lower levels than that of a 10-GHz magnetron. In a similar fashion, the klystron tube cavity must be greatly reduced in size in order to operate at 40 GHz, since the same mode of cavity resonance has to be used. Thus, the O-type device must similarly be operated at greatly reduced power in order to be "scaled" for use in the millimeter wave bands.

To illustrate this basic limitation on M-type and O-type tubes, Fig. 12-1 shows the pulse and CW power capabilities of these tube types versus frequency. A new type of tube also appears in that figure, the gyrotron tube. This tube is not limited by the design constraints of the conventional O and M types. The gyrotron is based upon the well-known resonance of the electron in a magnetic field, or the *gyro-resonance effect*. The resonators used in a gyrotron may be large compared to a wavelength, a condition permitting operating up to 200 GHz and higher.

The gyrotron can be built in many configurations. The simplest form is the gyromonotron oscillator that uses a single cavity. Other forms exist, however, that are similar in form to klystron and travelling-wave tubes. These are termed gyro-klystrons and gyro-TWT's.

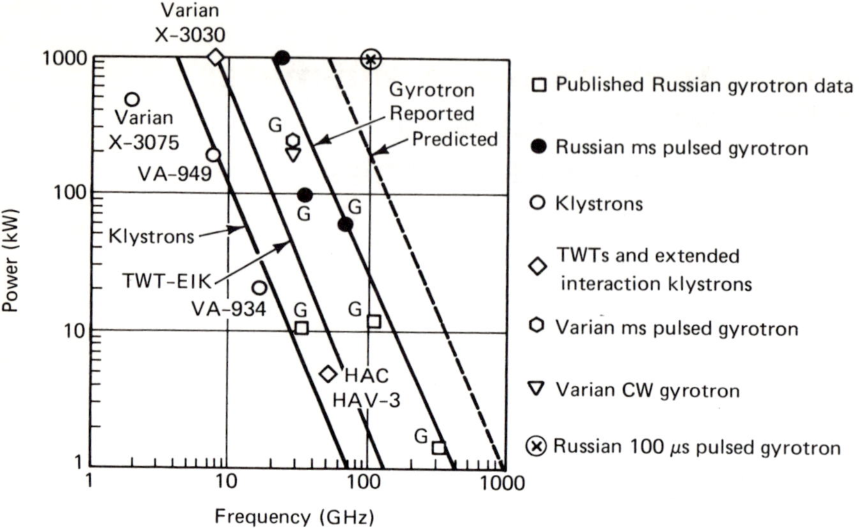

Figure 12-1: Microwave and Millimeter Wave Tube Performance

The gyrotron is sometimes called an electron cyclotron resonance maser. Chapter 10 considered maser and laser mechanisms and their corresponding devices. The gyrotron also employs a form of maser action to amplify waves, and thus uses a quantum mechanical effect to bring about amplification. Fortunately, we can use "classical" descriptions of the tube operation, making it unnecessary to go into quantum mechanics in order to understand how the device works.

The development of the gyrotron tube has made possible the availability of high-pulse power levels in the millimeter wave band. The tubes have recently been made available to U.S. research laboratories for use in the heating of plasmas for controlled nuclear fusion research. Out of this device development, many other applications can be forecast. Examples of other applications include high-resolution millimeter wave radars and high-directivity communications. The development of this tube will usher in a new series of electronic systems for worldwide commercial and military use.

12-2 THE GYROTRON PRINCIPLE

12-2.1 General Description of the Mechanism

The principle of the gyrotron was discovered independently by Twiss and Schneider of the U.S. and Gaponov of the Soviet Union, both in

1958. Laboratory confirmation was obtained about two years later in the Soviet Union. The mechanism depends upon a known characteristic of an electron moving in a magnetic field. When an electron moves parallel to a magnetic field, the field has no effect on the electron. That is, no force is exerted on the electron (the electron does not "see" the field). If the electron moves with a velocity v_0 that is not parallel to the magnetic field B, then forces will be applied to the electron by the magnetic field, and the tendency will be for the electron to move in a path in the form of a helix. The frequency of motion of the electron around the magnetic field is called the cyclotron frequency, and this frequency is proportional to the magnetic field strength. The theory predicted that there would be an energy interchange between the electron orbiting in the magnetic field and a nearby electromagnetic field at the cyclotron frequency. At very high electron velocities the electron would increase in mass, and the increase in mass would lower the cyclotron frequency. The interaction between the orbiting electron and the electromagnetic field would be such that if energy is given to the electromagnetic (EM) field, the electron would lose some mass and the phase of the cylcotron wave would be changed. The result would be a form of electron bunching analogous to the bunching action in a klystron tube. (See Chapter 9.) It is apparent, then, that the gyrotron is an extension of the earlier O-type tube principle. It has one significant advantage, however. The dimensions of the EM structure of the tube can be many wavelengths in diameter, since the resonance is due to the electron in the magnetic field, and not to a cavity structure or the dimensions of a helix. For this basic reason, the gyrotron interaction space can be much larger than that of a conventional O-type or M-type tube's interaction structure. Because of this basic fact, the gyrotron tube is capable of developing a much greater power in the millimeter wave band than the conventional O-type and M-type tubes.

It should also be noted that the gyrotron mechanism demands that the electron beam have most of its energy (i.e., velocity) in a plane perpendicular to the applied magnetic field, for reasons of efficiency. Clearly, if the interaction between the electron and the EM field is due to an electron's motion perpendicular to the magnetic field, then the electron's energy must be concentrated in that interaction mode. In order to fully utilize the gyrotron mechanism, the electron stream moving down the tube has to have most of its energy perpendicular to the stream flow. For this reason, most of the early development effort concentrated on the development of a special "magnetron injection gun" that would create a special electron beam that would have most of its energy perpendicular to the beam flow direction. The beam had to be hollow, also, in most cases.

In summary, the gyrotron mechanism permits much higher power levels at millimeter wave frequencies, but also demands a high velocity (semirelativistic) electron beam that has most of its energy in a plane perpendicular to the beam flow axis. This required the development of a form of electron gun unfamiliar to O-type and M-type tube designers.

12–2.2 Detailed Description of the Bunching Mechanism

The mechanism for the bunching of the electron beam in a gyrotron depends upon the orbiting of the electron around a magnetic field and the interaction with the rf field of the cavity that produces bunching. Velocity modulation occurs in the process, but it is in the form of angular-velocity modulation. The fundamental angular frequency ω of the electron is given by

$$\omega = \frac{d\theta}{dt} \simeq \frac{eB_0}{m_0\,(1\,+\,V/V_n)}\,,$$

where

$$e = \text{electron charge}$$

$$B_0 = \text{magnetic field}$$

$$m_0 = \text{electron rest mass}$$

$$V = \text{electron equivalent voltage.}$$

Figure 12-2 shows an electron spinning around the magnetic field $\bar{B}_0$ which is directed into the page. At position ①, the electron is accelerated tangentially by the electric field E. At position ③, the same electron is again accelerated, since the E field will have reversed by the time the electron reaches position ③, which is a half-orbit (180°). At positions ② and ④ of the figure, the electron does not receive tangential acceleration, but rather, receives radial acceleration. This has a smaller effect on electron motion, and thus can be neglected. The electron that receives the tangential acceleration will have increased equivalent voltage V, and thus

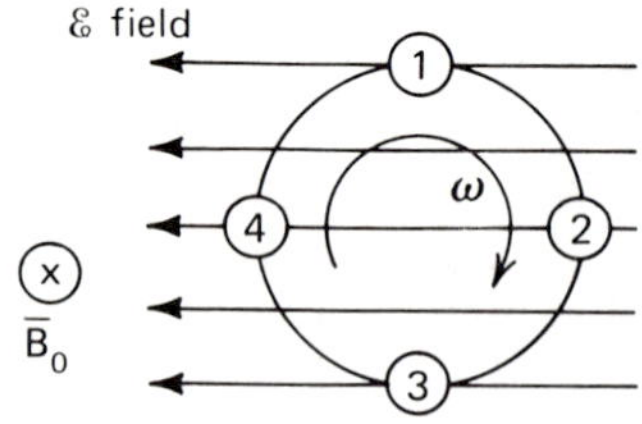

Figure 12-2: Field Interactions

will orbit with decreased angular velocity. Similarly, electrons that are tangentially decelerated will orbit with increased angular velocity. This is a form of positive feedback. When the bunches deliver energy to the field, they lose potential V and increase in angular velocity.

The microwave-induced angular-velocity modulation produces electron angular bunching as the beam is allowed to drift, similarly to the bunching process in the klystron tube. After moving through the drift space, the electrons will be bunched in angle, due to their initial angular-velocity modulation. These bunched electrons will give up a large fraction of their transverse energy to a properly phased microwave field. Figure 12-3 shows two phases of interaction between idealized electron bunches and a TE_{01} mode of a circular waveguide.

In Fig. 12-3a the electric field at the phase $t = t_0$ is in synchronism with the electron bunches, as viewed in the figure. In Fig. 12-3b, when the electrical phase had advanced by a half period ($\omega t = \omega t_0 + \pi$), the electric field is again in synchronism with the electron bunches, as viewed in the figure. Note that the electron orbit reverses by 180°, in synchronism with the electric field reversal. The electric field of Fig. 12-3 can be produced by the azimuthal electric field of the TE_{01} mode of a cylindrical waveguide.

Note that in Fig. 12-3 the orbital radius of the electron is r_1 and the average radius of the annular electron beam is r_g. A typical value for the TE_{01} waveguide radius is $2r_g$. r_g can be much larger than r_1, as indicated in the figure.

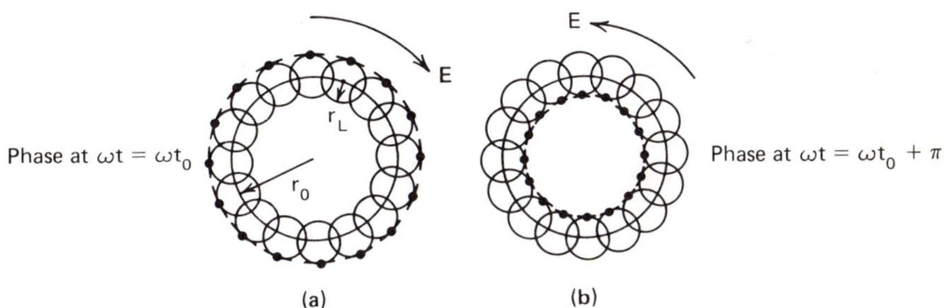

Figure 12-3: Idealized Electron Bunches

12–2.3 The Basic Types of Gyrotrons

The gyrotron tube can be described in terms of several generic types. Table 12-1, taken from the Soviet work, indicates the basic configurations that are possible. As the table shows, the basic oscillator is the gyro-monotron, which contains a single cavity. The electrons travel in a helical

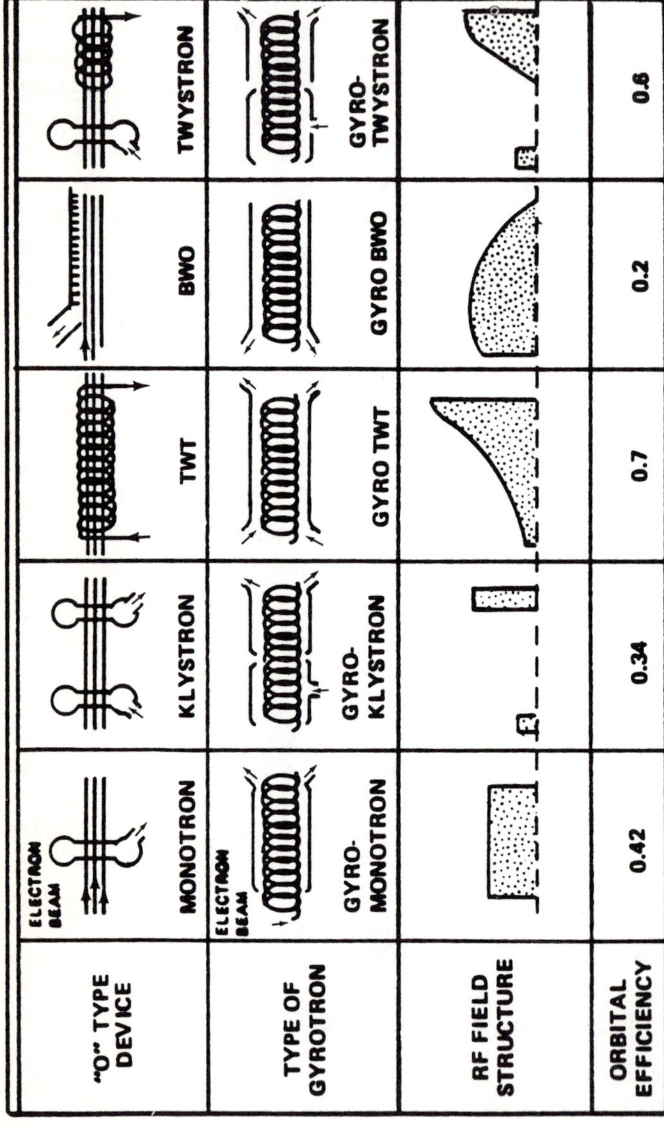

Table 12-1: Gyrotron Designs vs. Comparative O-type Devices *

*V.A. Flyagin et al., *The Gyrotron*, IEEE Trans. MTT-25, pg. 514–527, June 1977, © 1977 by IEEE. Reprinted by permission of IEEE.

path but interact with the fields near the cavity walls, thus necessitating a hollow electron beam. The equivalent O-type device would be a single-cavity monotron. (The closest to an equivalent O-type device would be the relfex klystron, as it has a single resonant cavity.) In all of the tubes in Table 12-1 a horizontal magnetic field is assumed, as in all O-type devices. The first row of the table is an attempt to show the correspondence to the more familiar O-type microwave tubes. Note that these gyrotron forms are derived from the familiar klystron, TWT, and BWO, and the not-so-familiar Twystron. In every case the electron-EM wave interaction region is replaced by either a fast-wave structure or a non-conventional cavity.[1] The row entitled "gyrotron rf field structure" shows the comparative electric field buildups in the various gyrotron types. The efficiencies of these new devices can be relatively high. For example, the gyro-TWT, the highest, has an estimated orbital efficiency of 70%. It is understood that the orbital efficiency is the conversion of electron orbital energy into rf field energy in the cavity. Cavity losses are not included in the orbital efficiency figure.

The five gyrotron forms of Table 12-1 have the basic advantages of their O-type counterparts, as will be seen in the sections that follow.

12–2.4 Comparison with Conventional Linear-Beam Counterparts

Using the cyclotron resonance interaction described in Sect. 12–2.3, a resonant cavity gyro-klystron amplifier may be built. Figure 12-4 below shows a U.S. experimental gyro-klystron that has been designed based on this principle. There are many similarities to the pulsed monotron oscillator. In fact, the same electron gun and collector were used to develop the gyro-klystron. Power gains of up to 40 dB have been achieved in this test device, with a peak output power of up to 50 kw at a frequency of 28 GHz. The similarity to a two-cavity klystron is clear. (See Fig. 9-2.) Note, however, that extraction of rf energy is not simple for a gyro-klystron. The rf energy must be propagated out of the last cavity and into the desired load, and yet the electron beam must also be collected and its power dissipated.

The problem is resolved by the transition section between the last cavity (the output cavity (6) of Fig. 12-4) and a larger-diameter circular waveguide output section. This connects to the beam collector section (8) of the figure. An output waveguide window (7) permits the rf energy

---

[1]All of the cavities are open-ended designs.

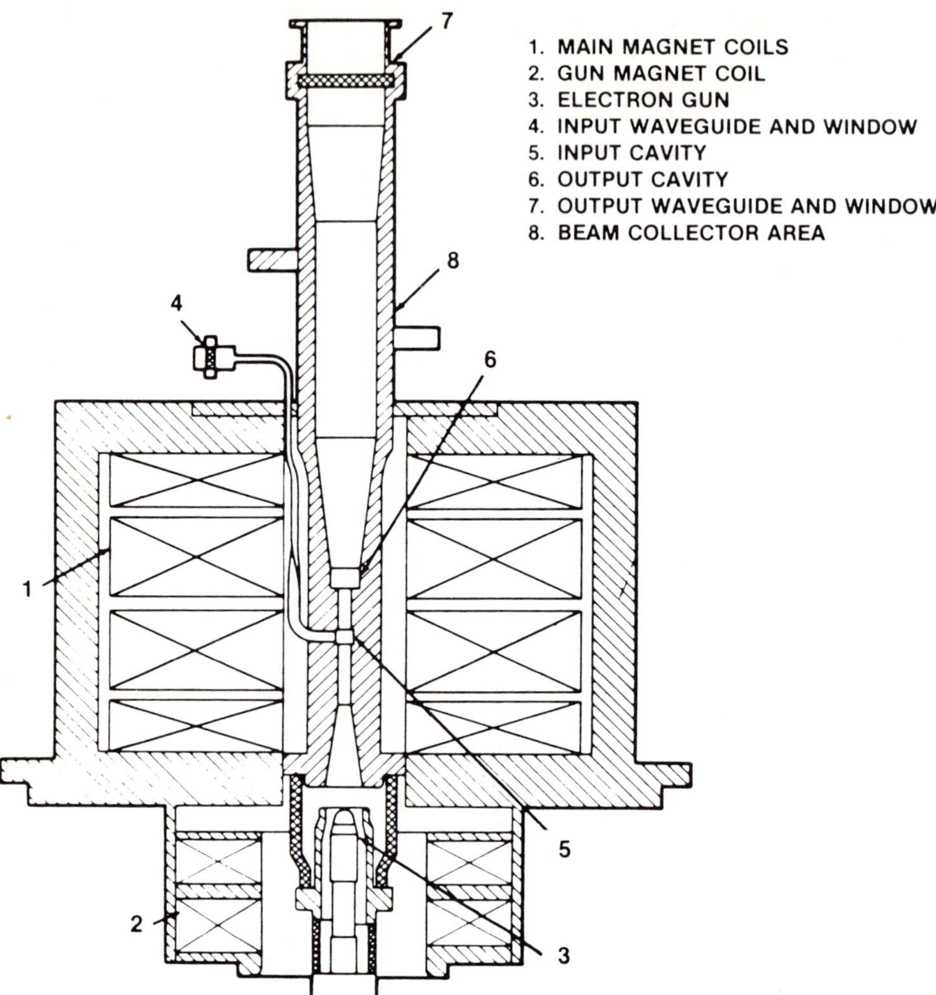

Figure 12-4: Gyro-Klystron Amplifier

to propagate outward while the tube vacuum is maintained. The output power is now in the form of a circular waveguide mode that usually has to be converted into a rectangular waveguide mode. This mode conversion is usually performed by means of a specially designed waveguide section added to the output. After mode conversion, conventional antennas can be driven by the gyro-klystron.

The initial performance of the gyro-klystron of Fig. 12-4 is encouraging. The design of the interaction space between the input and output cavities (located between area 5 and 6 of the figure) has apparently limited the efficiency to 10%, but, with redesign, efficiencies closer to the orbital value* of 34% indicated in Table 12-1 may be achieved.

The peak output power of 65 kw at 28 GHz is somewhat better than that of the conventional klystron of Fig. 12-1, but is not comparable to that of the Soviet gyro-monotron. The Soviet designers produced 1-MW pulses of about the same frequency, using millisecond pulses. To make a fair comparison, it should be realized that the amplifier gyrotron demands more careful control of cavity modes than the oscillator monotron. Once this is achieved, higher power gyro-klystrons will be produced.

The gyro-TWT concept of Table 12-1 has also been explored by U.S. experimenters. The gyro-TWT is a coherent amplifier that is designed to have much more bandwidth than a gyro-klystron. The gyro-TWT is formed by using a nonresonant "travelling-wave" structure in the interaction space. The travelling-wave tube of Sect. 9.3 also uses a nonresonant structure in a similar fashion. A schematic of the U.S. gyro-TWT is given in Fig. 12-5. The electron gun and the electron collector are similar to that of the gyro-monotron. In the U.S. design, the interaction waveguide is excited by two rectangular-to-circular waveguide transitions spaced 90° to each other. These two waveguides match into the two orthogonal circular waveguide modes that exist in the circular waveguide. The rf output is extracted from a window at the end of the collector section.

As in the case of the gyro-klystron, the output of this tube is in a circular waveguide mode, a mode not useful for driving an antenna. Again, a waveguide mode converter is needed for the transition to conventional rectangular waveguide. Now, since the gyro-TWT operates over a wide bandwidth, the mode converter must also perform well over the same bandwidth. A more advanced converter design is thus needed.

The experimental gyro-TWT of Fig. 12-5 yielded an initial effciency of 26%. This should be compared to the orbital efficiency of Table 12-1. Clearly, considerable development is required, since the orbital value is 70%. The design of Fig. 12-5 is not optimized, but with further development efforts, higher efficiencies and output powers can be anticipated.

The gyro-BWO of Table 12-1 can be constructed using a "backward-wave" mode (see Sect. 9.7) in a circular waveguide or similar structure. The structure would provide an electric field that would follow the angular bunching of the gyrotron but would have a growing wave action opposite to that of the axial electron-beam motion. A gyro-BWO would appear to

---

*The Soviet experimenters report "plate efficiencies" close to the orbital efficiency predictions; thus, the efficiency values of Table 12-1 can be approached in practice.

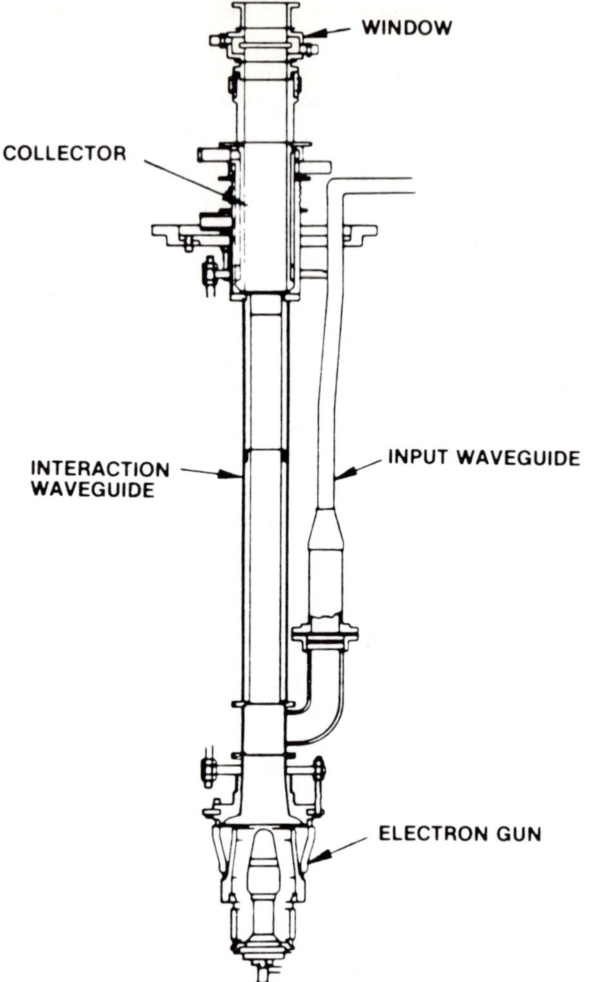

Figure 12-5: Gyro-TWT Amplifier

have low orbital efficiency, according to Table 12-1. For that reason it will not be considered further.

The gyro-twystron of Table 12-1 would be similar to the preciously considered gyro-klystron and gyro-TWT concepts. An interaction space must be provided between the cavity and the TWO interaction region to permit growth of the angular modulated bunches. It is expected that the overall orbital efficiency will lie between that of the gyro-TWT and the gyro-klystron. The orbital efficiency of the gyro-TWT of Table 12-1 lies between the gyro-TWT and the gyro-klystron. By analogy with the twystron of Fig. 9-13, it is possible that a gyro-twystron would possess a performance better than either the gyro-TWT or gyro-klystron separately. If so, as an amplifier it would possess high-power output capabilities over a wide bandwidth. This has yet to be proven by experiments. If successful, this would be a significant millimeter wave amplifier advancement.

12–2.5 Relativistic Effects in the Interaction Space

In Sect. 12–2.2, the angular frequency of the electron in a gyrotron tube was shown to depend upon the equivalent voltage V of the electron. In this section some of the physics of the electron in the magnetic field will be presented. The point here is that the electron in the gyrotron "sees" some relativistic effects that the O-type and M-type electrons do not. In fact, the gyrotron tube will not function without these relativistic effects. Furthermore, in Chapter 13 it will be made clear that some future very-high-power tubes will operate using much stronger relativistic effects. These must be considered in order to understand how these tubes operate.

The angular frequency of the electron in a gyrotron is given by

$$\omega = \frac{e\,B_0}{m_e},$$

where

$$B_0 = \text{static magnetic field}$$

$$e = \text{electron's charge}$$

$$m_e = \text{equivalent electron mass.}$$

Now, the mass of an electron is not fixed unless it is at rest. Call this mass the *rest mass* m_0. When it is in motion, the electron appears to have the mass

$$m_e = \frac{m_0}{(1 - (v/c)^2)^{1/2}}$$

where v is the electron velocity and c is the speed of light. As the electron velocity approaches the light velocity c, m_e increases sharply. In other words, the electron *appears* to become very heavy (massive) as its speed nears that of light. The emphasis is placed on "appears," as this is the nature of relativity: as the electron is viewed as it goes by us, it looks to be very heavy. The effect of this on ω is that the angular frequency is reduced even though the field B_0 is unchanged. The faster the electron moves, the more the value of m_e is increased, and the lower ω becomes.

The above effect is a modulation of the electron gyro frequency with electron speed. Since a higher-speed electron has more energy (kinetic energy $\simeq 1/2\ mv^2$), the more energetic electron has a somewhat lower gyro frequency.

For the case of the gyrotron in this chapter, the effects are small, as the electron beam is not very near to the speed of light. It can be said that the beam is weakly relativistic. The effect, though, is that when the electron bunches give energy to the rf field in the gyrotron, they lose energy. But, in the process, they have an increase in angular rotation frequency due to having become less massive. This increase in angular modulation produces more bunching, and thus produces a feedback effect. The point is that the relativistic effects make possible the chain of events that result in the production of rf waves in the output guide via the angular bunching effect.

12–2.6 Operation at Harmonics of the Gyrofrequency

The frequency of operation of the gyrotron is given approximately by

$$\omega \simeq n\ \frac{e\ B_0}{m_0}\ (1 - \beta^2/2),$$

where $\beta = v/c$, n is an integer, and $\beta << 1$.

The "frequency" ω is really the electron's angular frequency. The electromagnetic (rf) frequency, in Hertz, is given by

$$f_0 = \frac{\omega}{2\pi}.$$

Thus, the gyrotron's frequency is given by

$$f_0 = \frac{n\ e\ B_0}{2\ \pi\ m_0}\ (1 - \beta^2/2).$$

n is simply a whole number, i.e., an integer. n equals two signifies that the rf frequency of the gyrotron is chosen to be twice (at the second harmonic of) the basic electron's gyrofrequency. Gyrotrons have been operated at harmonics of the basic gyrofrequency with success. Where possible, this is desirable, as much less magnetic field strength B_0 is then needed.

The Soviet literature shows that the orbital efficiency of a gyro-monotron can be high even at higher harmonic values of n. Table 12-2 below shows their predictions. The n = 1 value corresponds to the value given in Table 12-1 for the gyro-monotron.

There is the possibility in gyro-monotron operation to sacrifice efficiency for operation at lower field levels B_0. This is of value, as it may be possible to eliminate the need for a supercooled (superconducting) magnet at millimeter wave frequencies by lowering the required value of the magnetic field B_0 by the integer value n.

Harmonic Value (n)	1	2	3	4	5
Orbital Efficiency	0.42	0.30	0.22	0.17	0.14

Table 12-2: Harmonic Gyrotron Efficiency

12–3 GYROTRON OPERATION PRINCIPLES

This section will consider the principles of electrical operation of the gyrotron, based upon the experience of U.S. high-power microwave tube developers. The requirements for operation of a gyrotron, whether an oscillator or an amplifier, are similar to those for the development of any linear-beam (O-type) tube, such as a high-power klystron or TWT. Individual differences will develop, depending upon specialized requirements, but the basic concepts of operation are not expected to vary significantly.

Figure 12-6 is a schematic diagram for operating a basic gyrotron.

The body of the tube is operated at ground potential for personnel protection reasons. The beam voltage and gun anode voltages are referenced to the cathode. Not shown are the currents to the various magnet coils. For the case of the gyro-klystron (see Fig. 12-4), the currents in the main magnet coils are controlled separately. Typically, 26 volts per coil and up to 500 A per coil are employed for the magnet energization. In Fig. 12-4, two gun magnet coils are used with 33 volts applied, with a current level of 10 A. Typically, the electron gun and gun coils are oil cooled; the main magnet coils are water cooled, with a typical flow of 15 gpm.

For typical operation of the gyrotron in Fig. 12-6, beam voltage could reach 80 kV, with a beam curent level of 8 A. The gun anode could require about 25 kV. The heater voltage and current are adjusted for temperature-limited operation. In other words, the beam current is con-

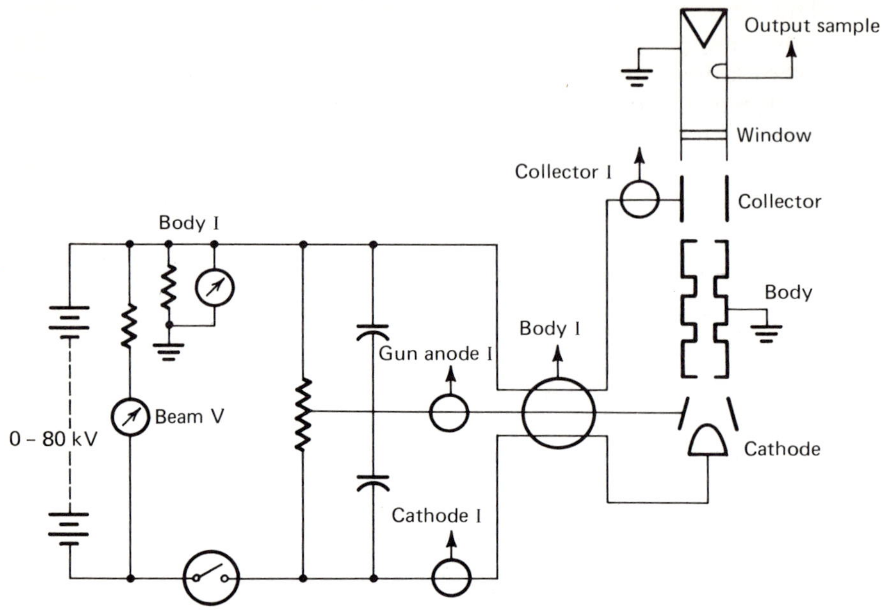

Figure 12-6: Gyrotron Operating Connections

trolled by changing cathode emission via its temperature. Varying the heater voltage will provide this control.

The sensitivity to the above critical parameters is given in Table 12-3. Referring again to Fig. 12-6, we see that the collector will be far enough away from ground potential to permit monitoring the body current. In the event of an interal (or external) tube arc, special "crowbar" protection circuits are provided that remove the high potentials in less than 5 μs, thus preventing damage due to excessive energy discharge into the tube.

Parameter	Typical Value	Sensitivity (dB %)
Gun Coil	10 A	0.11
Heater Voltage	12 V	0.32
Gun Anode Voltage	25 kV	0.39
Beam Voltage	80 kV	0.54
Main Magnet	500 A	1.21

Table 12-3: Sensitivities

Protection is also provided against failure of the magnetic field, coolant flow in tube or magnet, and arcing failures in the output waveguide. The latter failure protection requires an arc detector that looks directly at the output window of the tube. If the tube is pulsed, some protection must also be provided against both long pulses and undesired CW operation. Often, a small vacuum pump is used to maintain minimum vacuum levels (under 10^{-8} Torr). It is usually tied into the protection circuitry to remove beam voltage in the case of loss of vacuum.

12–4 GYROTRON APPLICATIONS

12–4.1 Commercial Potential

One motivation for the development of the gyrotron tube is the need for high-power millimeter wave energy in controlled nuclear fusion. As the higher plasma densities and temperatures are reached, high energy in the rf form is needed to increase the thermal energy of the fusion plasma. The frequencies used must be near the natural frequency of the plasma, since it is at this frequency that it can absorb energy. For the plasma densities considered, this frequency is in the millimeter wave bands. The very-high-power pulsed gyrotron of Fig. 12-1 at 100 GHz is probably for use in the Russian Tokomak plasma experiments. These experiments have nuclear fusion as their objectives.

The U.S. Oak Ridge National Laboratory is presently using 28-GHz, 200-kW gyrotrons* in their plasma fusion experiments. In particular, these tubes are being used on the ELMO BUMPY Torus at Oak Ridge.

Other commercial applications could include airborne radar for traffic control and for advanced communications use. The millimeter wave band permits very high data rates, due to the high frequencies involved. The high power levels generated by gyrotrons will open new communication bands, particularly within the low-loss propagation "windows" of the atmosphere. Very-narrow-beam (high resolution) antennas are possible at these frequencies, making possible the focusing of the energy into very small volumes. Guarded or secure communications become possible over these narrow beams of EM energy.

Industrial applications will also follow. Some industrial processes could be advanced by the availability of very-high-power millimeter wave energy. It is not possible at this time to fully explore the possibilities inherent in this recent device development.

---

*A CW power level of 212 KW was achieved by these tubes.

The achievement of nuclear fusion will in itself have high commercial applications. Such use may greatly expand the use of the gyrotron tube, resulting in eventual off-the-shelf availability.

12–4.2 Military Applications

A strong possible future application for the gyrotron is in military radar systems. This application will require coherent amplifiers, preferably with large bandwidth capability. As seen in Sect. 12–2.4, a gyro-klystron or gyro-TWT will require more development effort. The present monotrons have had high promise, and in a few years will be available for general use. A parallel development of coherent gyrotron amplifiers will be required before use in military systems is likely.

The promise of high resolution, compact design, high bandwidths, and high transmittal power make the development of millimeter wave radars based upon the gyrotron concept quite likely.

12–4.3 Propagation Limitations

Use of the gyrotron in applications involving wave propagation in the atmosphere (radar, communications, etc.) will be limited by the attenuation of the atmosphere. At some frequencies millimeter waves are very strongly attenuated over short path lengths. 60-GHz waves at sea level are attenuated at the rate of roughly 14 dB per km. There are propagation "windows," however, where attentuation losses are very low. Such bands will be the most desirable for radar or communication use.

The susceptibility of a millimeter wave system to excess attenuation due to rain and fog is also a consideration. At very high millimeter wave frequencies, the atmosphere becomes opaque.

In short, attenuation losses in the atmosphere over much of the millimeter wave band will restrict use to certain preferred bands, or windows. Presently, these windows are fairly well defined. With the availability of very high gyrotron power levels, however, these system losses lose some of their impact. The seriousness of the loss must be compared with the availability of rf power levels. 20 dB more power can offset a considerable system loss.

12–5 COMMERCIAL AVAILABILITY

12–5.1 Present

The U.S. microwave tube community has now started to deliver a few 28-GHz gyro-monotrons to the R-and-D community. In a few years 90–GHz gyrotrons will start to become available. At the present time the monotron is just becoming available to the U.S. R and D community. The gyro-TWT is now under development, as is the gyro-klystron. These may become available in three to five years, depending upon the level of research expended on them.

12–5.2 Future

The very high promise of the gyrotron concept, plus the recent evidence of its successful development in the monotron version, suggests that it will be successfully developed in both its oscillator and amplifier versions. This will lead to commercial availability of both the oscillator and the amplifier versions, further spurring on gyrotron research and practical development.

13   Intense Relativistic Beam Devices (IREBs)

13–1 INTRODUCTION

In previous chapters, various forms of microwave and millimeter wave electronic devices have been studied. The solid-state devices showed the promise of generation of devices in the microwave and millimeter wave frequency regions that could serve as oscillators and amplifiers of EM signals, but all at relatively low or moderate signal levels. The Josephson junction device was shown to detect extremely low-level signals at extremely high frequencies. The chapter on the electron cyclotron maser showed some new vistas for power generation and amplification at millimeter wave frequencies (30-300 GHz). In this chapter consideration is given to the step beyond the conventional gyrotron into the field of millimeter wave pulse power generation using electron beams whose velocity is very near the velocity of light. Such devices, now being studied in laboratories around the world, use electron beams of extremely high energy to generate EM waves ranging from a few GHz to over 100 GHz. Perhaps they could supply the final link to controlled nuclear fusion. They apply, however, all of the principles discussed to date. The reader would do well to absorb the material of Chapters 8, 9, 10, and 12 before reading this chapter. The IREB device is built upon those principles.

13–2 THE IREB DEVICE PRINCIPLES

13–2.1 Extended Conventional Devices

Intense relativistic electron beam (IREB) research has been conducted since the late 1960's in the generation of flash X-ray devices. The energies in the beams led to the exploitation of ultrahigh-power microwave and millimeter wave power generation techniques. They are developing toward special electron laser devices that could fill the gap in power generation between the optical laser and the conventional microwave tube. Consider first the extension of conventional microwave tubes into the IREB domain.

13–2.1.1 Relativistic magnetrons

The conventional magnetron of Chapter 8 has been extended into the IREB domain by ingenious redesign to permit the injection of an electron beam with relativistic energy. Experiments at the Naval Research Laboratory (NRL) in 1976 showed that pulse powers of 4 GW with a pulse width of 50 ns could be generated at a frequency of 3 GHz. This work showed that the magnetron could be scaled for use with IREBs. It also demonstrated that the relativistic effects did not distort the electron geometry in the device and that pulse widths of up to 1 μs were possible. Work at the Massachusetts Institute of Technology (MIT) in 1976 using an IREB magnetron operating at 3 GHZ in the Pi-mode showed that with a cylindrical geometry and a central cathode, a peak power of 1.7 GW could be generated at an efficiency of 35%. More recent work at NRL (1979) showed that a 0.5-GW, 20-ns pulse could be developed at 3.2 GHz with an efficiency of 20%, using a 54-vane hybrid inverted coaxial magnetron. This magnetron used a more complex (higher-order) cavity mode with a circular wave-guide mode (TE_{01}) output. These experiments were followed by Russian research in which similar peak powers were generated, but at near-microsecond pulse widths.

This research proved that the magnetron could generate gigawatt pulse powers at frequencies under 10 GHz and pulse widths of the order of 1 μs. The efficiencies generally did not exceed 30%, but this seemed to be typical of an IREB device.

13–2.1.2 Linear beam (O-type) devices

The O-type device can also be driven by an IREB source to achieve phenomenal pulse power levels. In the U.S. in 1975 (NRL) a multiple-cavity klystron developed 0.6 GW of power at 3 GHz with a 50 ns pulse. The efficiency was 20%. It was a special device using an annular beam and a circuit of 20 stainless steel resonators. Also in the U.S. (Cornell University) an IREB was used to drive a BWO to generate a 0.5-Gw pulse at 9 and 11 GHz with a pulse width of 70 ns. The efficiency was 17% for these tests. The BWO used a rippled-wall circular waveguide and a 1-cm diameter electron beam. The design used a special slow-wave microwave structure design in the circular waveguide.

The ultimate limit of rf breakdown was probably achieved in the above klystron experiments and, therefore, it can be assumed that a few GW is the upper limit for either the O-type or M-type device operated in the IREB domain. The frequency of operation of any of these devices in the IREB mode is certainly limited to 10 GHz.

13–2.2 Nonconventional Devices

13–2.2.1 Introduction

The term 'nonconventional devices' refers to the new devices that are unique to IREB designs. These devices can also be called fast-wave designs, since, in general, the EM wave phase velocity is greater than the speed of light. (An EM wave can have a phase velocity greater than that of light, as the information in the wave must travel at the group velocity, which is slower than the speed of light. This is explained in simple terms in Sect. 8–5.) The basic design of a fast-wave device is given in Fig. 13-1.

13–2.2.2 The IREB CRM (gyrotron)

The gyrotron has been described as a cyclotron resonance maser (CRM) in Chapter 12. The IREB CRM did not develop as an extension of the thermionic cathode gyrotron of Chapter 12. The earlier IREB experiments showed that the main mechanism for rf radiation was the CRM mechanism. In the U.S. experiments (NRL) on ubitrons (using a rippled magnetic field in the interaction region), greatly enhanced radiation appeared at the cyclotron frequency ω. Also, in the Soviet plasma heating experiments using IREBs with substantial transverse energies, large cyclotron emissions were discovered. Subsequent experiments have shown,

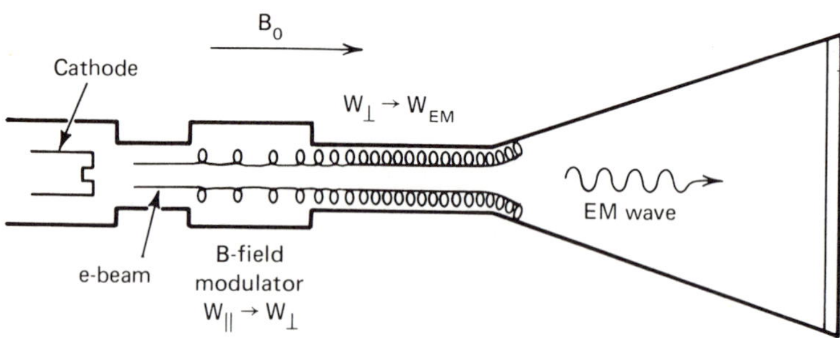

Figure 13-1: Basic Fast-Wave Device Design

in both countries, that the CRM mechanism can be useful for IREB microwave generation and amplification.

In U.S. experiments (NRL) performed in 1975, experimental powers of 1 GW at 7.5 GHz and 2 MW at 75 GHz were generated. The efficiencies were under 1% for the above experiments. The Soviets have performed more recent work and have claimed 2 GW at 3 GHz, but with 30% efficiency. The recent Soviet success is no doubt due to a better electron-beam energy distribution, putting most of the beam energy in the desired transverse direction. There is promise for ultrahigh power at shorter wavelengths in this work.

13–2.2.3 The free-electron laser (FEL)

The free-electron laser (FEL) uses a wiggling of the electron beam by some means to provide interaction with the EM wave. In the earlier U.S. ubitron work a periodic magnetic field was used to pump (undulate) the electron beam, using what was termed "stimulated scattering." This was the earliest demonstration of this interaction mechanism to achieve notable success. In that work (in 1960) ubitron amplifiers were built at 2.5 GHz (170 kv) and 54 GHz (80 kv) which had peak output powers of 1.2 MW and 150 kW with 10% and 6% efficiencies, respectively. Free-electron lasers can be built to operate throughout the millimeter wave spectrum and up to optical wavelengths.

In the 1977 work of NRL, an FEL was operated at 750 GHz with an output power of 1 MW in a 60 ns pulse. An intense 2-cm pump wave was generated by a CRM mechanism, using the same beam. An effective two-stage pump was developed, using the same electron beam. The efficiency, however, was under 1%.

In the same year, Columbia University developed an FEL using a rippled magnetic field that generated a 15-ns, 8-MW pulse at 150-200 GHz. The efficiency achieved was about 2%.

13–2.2.4 The reflex triode

This device uses a modification of a simple triode configuration to generate gigawatt levels of EM energy at X-band (8-10 GHz). In experimental work at NRL in 1977, high-power microwave radiation was observed when a positive pulse was applied to an anode (grid) of a reflex triode. A 100-MW pulse of radiation at X-band was observed, with an efficiency of 1.5%. The same mechanism for EM generation was used to explain a French experiment in which over 1 GW was generated at X-band.

Recent (1978) work at the U.S. Harry Diamond Laboratories (HDL) resulted in gigawatt-level powers at X-band using a relativistic beam reflex triode. Single microwave bursts of 10-ns duration were achieved, with a conversion efficiency of 1 to 2%. The triode geometry is given in Fig. 13-2.

In the HDL work, a magnetic field of 0.1 to 0.4 Tesla was applied in the y-direction, and a pulse of 1-2 Mv was applied between anode and cathode. Explosive electron emission occurred, resulting in an average current of 15-20 kA through the triode in the form of 15-ns pulse. Relativistic space charge oscillations occurred that had a rich frequency content. The triode could be tuned by varying the grid-to-anode gap, the gap voltage, and the magnetic field B_0.

The reflex triode has also been investigated by A.N. Didenko of the U.S.S.R. in a configuration similar to that of the NRL experiment. Didenko described a virtual cathode as forming near the plate of Fig. 13-2, with a space charge potential well forming near the grid (anode). Didenko generated a peak power of 1.4 GW at a wavelength of 9 cm (3.3 GHz) and a pulse duration of 40 ns. He quoted an efficiency of 12% using a 450-kV pulse and a current of 25 kA. He found that the output wavelength could be tuned from 2 to 6 GHz by varying the anode voltage, in agreement with NRL work.

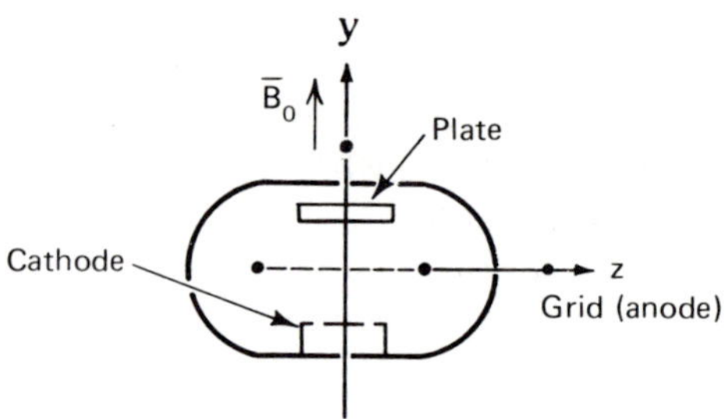

Figure 13-2: Reflex Triode Geometry

13–3 STATE OF THE ART AND THE EFFECT OF POWER CONDITIONING

The IREB device is in its early stages of development. Very-high-energy electron beams are required for its operation, and thus, very-high-drive voltage sources are needed. The IREB device rarely exceeds 10% in efficiency, so a considerable loss of energy is experienced in its operation. However, it can provide gigawatt levels of power from X-band up to about 20 GHz. 10-MW pulses have been generated at 40 and 200 GHz using a BWO and an FEL, respectively.

The power conditioning necessary to produce the electron beam is described in Fig. 13-3. The output electron beam could then be used in extended conventional devices (magnetrons, etc.) or in nonconventional devices. For the latter, Fig. 13-1 gives the basic fast-wave device configuration using the pulsed electron beam.

The IREB device thus requires special power supply "conditioning" to form the intense voltage pulse and current capability needed to generate gigawatt-level pulses. At a frequency of 200 GHz and an efficiency of 2%, a dc pulse of 0.5 GW is needed to generate a 10-MW pulse. This means that large installations are needed, due to the low efficiency involved.

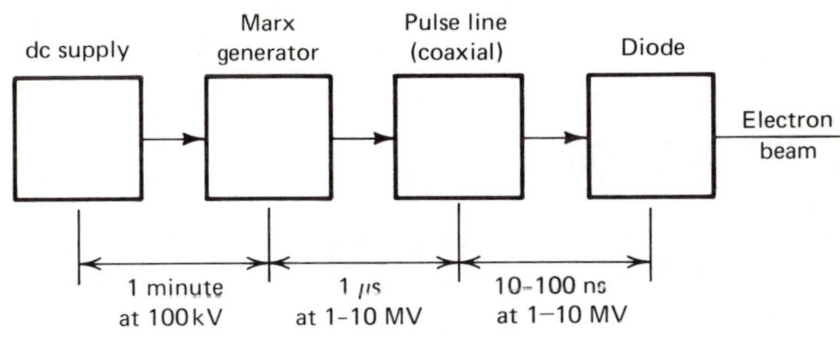

Figure 13-3: IREB Generator Design

13–4 IREB APPLICATIONS

13–4.1 Research Use

A principle use for IREB devices is in magnetic-confinement fusion experiments. Particle accelerators driven by very-high-power rf pulses could benefit greatly by the gigawatt-level pulses possible via IREB devices. Another research field that may employ these devices is the field of isotope separation. The frequency ranges needed for these research applications could extend from the conventional microwave bands up to 1000 GHz (into the millimeter and submillimeter bands).

13–4.2 Commercial Applications

With the advent of efficient IREB device systems that can be made into some reasonably sized installations, commercial use in large installations may be possible. Compact devices cannot be foreseen at this time.

14 Device Concepts for the Future

14–1 INTRODUCTION

In the previous chapters, a broad range of microwave and millimeter wave devices and device principles have been discussed. The emphasis has been placed upon the device principles, not upon hardware development. There are usually a number of hardware approaches to achieve the device implementation, so the latter *can* be flexible. An example is the achievement of the multiple cavity anode structure of the magnetron oscillator. Many possibilities exist in addition to the slotted cyclinder anode configuration of Fig. 8-8, as illustrated by Fig. 8-11.

With this emphasis on device principles in mind, this chapter is devoted to a discussion of the device principles of the future, the principles upon which future hardware may well be developed. The section headings of this chapter are chosen to emphasize those new principles that can lead to new and improved devices for microwave and millimeter use. Some new material will be introduced, material that has not been developed in the previous chapters. This approach, however, will give more liberty to explore concepts that at the present writing have not been put into practice or become mature in the state–of–the–art.

Projections in the device art are rather rare. However, such projections will be made here despite the difficulties inherent in making them. The basic principle is often overlooked. As someone once said, "Before you make the forecast, be sure to read the morning paper." Forecasts must be flexible.

Allowing for this uncertainty, the best approach is to provide an audit trail of the logic of the forecast so that it will at least be clear how the forecast was derived. Forecasts without audit trails belong to the true prophets alone.

Projections are based upon the wisdom of the past. Few devices are developed using only new principles. Existing mechanisms are often improved. The strapping of the anodes of a magnetron is an example of this evolution (see Fig. 8-8 and its accompanying text). Projections are not difficult to make since new means will always be found that force the device to conform closer to its ideal model or principle.

But projections must also allow for quantum steps in performance. When a completely new principle is discovered, such as Brian Josephson's (see Chapter 11), a whole new series of device concepts is envisioned. Sometimes the inventor can imagine possibilities by looking at the theory alone. The quantum step of the electron cyclotron maser (gyrotron) is an example. No Nobel prize was awarded for the gyrotron discovery, but it *was* the result of new principles applied to known device limitations. Completely new cathode and resonator designs were needed. But in this latter example, much in the way of conventional microwave technology was adapted to permit the desired orbital electron/EM coupling in the gyrotron interaction space. The more basic discovery was that of Josephson, who also predicted basic device capabilities before laboratory confirmations were performed.

The devices of the future will be either evolutionary or revolutionary. The revolutionary concept is based on new concepts which do not lend themselves to forecasting techniques. With this in mind, the following material is meant to be evolutionary. The lessons of the past do not have to be relived, and the technical discoveries of the past will not be overlooked.

14–2 **THE EXTENDED VACUUM DEVICE**

The gyrotron family of devices of Table 12-1 is taken from the foreign literature, even though it was printed in a U.S. publication. The gryo–TWT, for example, is a travelling wave tube that uses fast wave rf coupling in contrast to the slow wave coupling to the electron stream used in the conventional microwave TWT. In a sense, the conventional TWT is extended into a higher frequency regime by the new design concept. The device is extended into the millimeter wave regime by the use of the electron cyclotron maser principles covered in Chapter 12. This is a semi–evolutionary change in device design as the electron cyclotron maser was a breakthrough when it was discovered in 1958.

Since conventional microwave tubes have apparent counterparts in the gyrotron family (as viewed in Table 12-1), are they limited in their development? The foreign press says *no*. Other possibilities exist. The efficiencies listed in Table 12-1 are for linear magnetic fields and smooth, non-optimized resonators. Foreign researchers believe that a careful optimization of the resonator contours and a "tailoring" of the longitudinal magnetic fields could greatly improve orbital efficiencies or the efficiency of converting electron beam energy units into rf output energy. Beyond these evolutionary improvements, other device concepts are possible.

All conventional gyrotrons have a symmetrical, hollow electron beam design. If a rectangular or slab design can be made, then electron gun designs simpler than those now in use could be made.

The vacuum gyrotron requires strong magnetic fields for its operation and thus uses super-conducting magnets for operation above 50–60 GHz. Harmonic operation can permit strong reductions in the required longitudinal magnetic field. Usually lower efficiency results from this choice (see Table 12-4). This is an apparent limitation. A quantum step would be to find a way to permit, for example, 95 GHz operation, at full efficiency, without the need for super–conducting magnets. This latter type of development will be explored since the supercooled magnet structures tend to detract from the tube's usefulness, particularly for mobile use.

14–3 SOLID STATE DEVICES

A possible future device for the millimeter and submillimeter wave bands is a form of gyrotron using semiconductor materials. Conventional Gunn diode and IMPATT devices suffer severe deterioration in their characteristics when operated in either millimeter or submillimeter bands. A semiconductor gyrotron would not need magnetic field strengths of a vacuum gyrotron, as the effective mass of an electron in solid state materials can be greatly reduced compared to the electron in free space. Since the electron cyclotron frequency is given by:

$$\omega_b = \frac{e\,B_0}{m^*},$$

where

$$m^* = \text{effective electron mass},$$

a greatly reduced value of m^* means less field B_0 is needed to achieve cyclotron resonance at millimeter wave frequencies. Calculations indicate that for Indium Antinomide

$$m^* \simeq 0.014 \, m_e \,,$$

where m_e is the free electron mass. This will permit some millimeter band operation with a few kilogauss of magnetic field, a level not requiring a superconducting magnet.

The conventional gyrotron depends upon the variation of mass of the electron with velocity:

$$m(\mathcal{V}) = \frac{m}{\sqrt{1 - (\mathcal{V}/c)^2}}$$

The solid state gyrotron could use a similar dispersion law for the conduction band of Indium Antinomide (In Sb) material:

$$m(\mathcal{V}) = \frac{m_0^*}{\sqrt{1 - \left(\dfrac{\mathcal{V}}{\mathcal{V}_g}\right)^2}} = \frac{0.014 \, m_e}{\sqrt{1 - \left(\dfrac{\mathcal{V}}{\mathcal{V}_g}\right)^2}}$$

where

m_e = free electron mass

$\mathcal{V}_g$ = electron velocity in In Sb due to band gap energy E_g, computed using m_0

m_0^* = effective electron mass

This equivalence to the vacuum gyrotron could permit small, efficient millimeter and submillimeter wave generators using conventional magnet structures.

The solid state gyrotron would probably be cooled to 77° K for proper operation, but this is far better than having to cool a large magnet structure to 4° K as in the vacuum equivalent.

The power output of such a device would not be large, but larger than today's solid state millimeter wave generators. As seen in Fig. 9-15, CW solid state generators at 60 GHz are limited to under 50mw. There are no semiconductor submillimeter generators available at this time.

14–4 THE FREE ELECTRON DEVICE (UBITRON)

The free electron laser (FEL) is a device for generating high power at wavelengths shorter than 1 mm. Its operating frequency is determined not by energy levels in atoms or molecules, but by the scattering of free electrons passing through a spatially varying magnetic field. The FEL may produce radiation at frequencies varying from about 10 GHz (30 mm) to the IR and optical frequencies. Figure 14-1 shows predicted FEL response compared to laser technology and the gyrotron. The device is called a *ubitron* in the Soviet Union.

A typical FEL is sketched in Fig. 14-2. Raizer in the U.S.S.R. reported the generation of 30Mw pulses with a ubitron of this design using an electron beam at energy 700 Kev, with current pulses of 1 KA. The electron beam is passed through a periodic magnetic field produced by a solenoid enclosing a system of copper rings. The copper rings cause a rippling magnetic field. The electron beam of Fig. 14-2 is bunched as it passes through the rippled magnetic field, and the oscillation of these bunches leads to a conversion of the electron transverse energy into EM energy. In the case of Raizer's work, an efficiency of 5% was achieved at a frequency of 10 GHz.

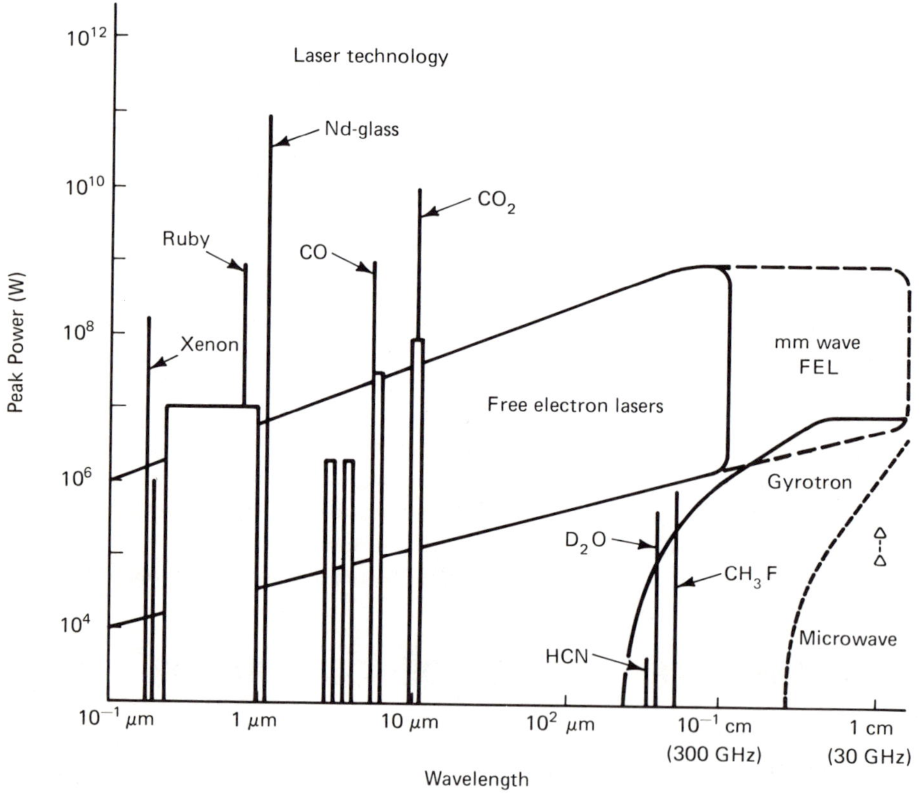

Figure 14-1: Power Capabilities of the Free Electron Laser

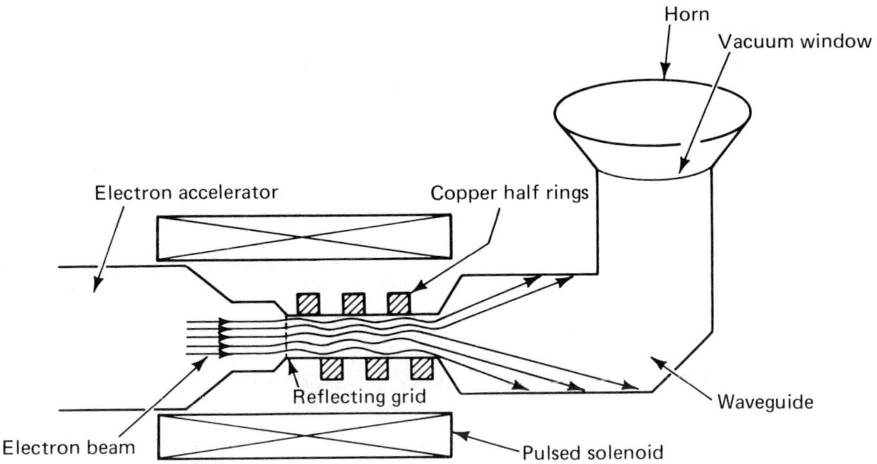

Figure 14-2: Relativistic Ubitron

Tkach of the U.S.S.R. built a relativistic ubitron with an interaction region under vacuum and filled with a plasma. As in the gyrotron case, the plasma reduced the beam velocity spread due to Coulomb repulsion forces, increasing efficiency and output power. He generated 20 Mw pulses at 11 GHz by this method.

Referring to Fig. 14-1, the above experimental FEL power levels are modest (20-30 Mw), but the ultimate capability at millimeter wave frequencies is believed to be gigawatt levels, as indicated by the figure.

14–5 SUPERCONDUCTING DEVICES

14–5.1 Josephson Detectors and Generators

As stated in Chapter 11, the discovery of Josephson has resulted in a potential family of devices based upon the two Josephson effects. With advances in cryostat technology, the need for cooling the detector/generator element to 4° K will become less of a problem. The junction itself is quite small, but the environment around it should be cooled as well to be compatible with the noise temperature of the junction.

When higher critical temperature superconductors are developed, the Josephson devices will become more popular due to the less stringent refrigerator requirements.

The linear relationship between the dc bias voltage and the Josephson junction oscillation frequency can be used for precise, wide–band frequency generation at millimeter and submillimeter frequencies. The power outputs will be low, but the frequency control mechanism is very simple. It is basically a voltage–controlled millimeter/submillimeter oscillator. Panoramic receivers over these frequency bands could be developed using the junction as a voltage–controlled local receiver oscillator.

As a detector, it can be used in conjunction with a bolometer to detect signals over very broad bandwidths, as described in section 11–6.5. With the achieved sensitivity of 10^{-15} watt/Hz$^{1/2}$ over the range of 0.1 to 1.0mm, it has high promise for millimeter wave and submillimeter wave detection. The slow response time of the bolometer element is a limitation.

14–5.2 Indium Antinomide (In Sb) Detectors

Indium antinomide (In Sb) detectors in the millimeter/submillimeter bands have demonstrated high detection sensitivities and have promise for detection up to IR frequencies. They require cooling to 4° K as in the Josephson devices, but have had the benefit of more extensive development. They can operate over wide bandwidths, the bandwidth being determined by the resonating structures and not the element per se. They have been flown in balloon–mounted radiometers and other higher altitude platforms.

Detection sensitivities (NEP levels) approaching 10^{-15} watt/Hz$^{1/2}$ have been approached by radiometers using these detectors. Such radiometers have been able to sense temperature changes of 10^{-2}K° in the submillimeter bands, making this a very impressive remote temperature sensor.

The development of reliable cryostatic systems for providing the 4° K cooling of the In Sb sensor will make this device more suitable for field use.

Such detectors may be suitable for use in communication systems between satellites. With no attentuating atmosphere between them, and with the use of modest power, directional beam, highly reliable communication links between space vehicles might be achieved. The satellites would have to provide the cryogenic systems to cool the In Sb detector elements. The outside temperature would vary considerably and would be a function of the sun's illumination, requiring careful control.

14–5.3 Cavity "Ringing" Devices

A superconducting cavity has very low loss. Such a resonator could be periodically pulsed and set to "ringing" electromagnetically. This is a very simple concept. The magnetron's cavity is set into ringing by the spoked electron stream moving across the entrance region of the cavity. A very high Q cavity would ring for a long time between impulses. For

CW operation the impulses would be synchronized with the cavity phase to permit driving to overcome cavity losses. The cavity dimensions and mode of oscillation would determine the oscillation frequency. For pulse operation, assuming no required coherence between pulsed output signals, there are fewer timing and synchronization problems.

When devices capable of switching at millimeter wave frequencies are developed, they could drive such superconducting cavity resonators. Presumably a Josephson junction could achieve this action since it has a switching time of 10^{-11}s. The power it could switch would be very small, however,

When precise frequency control is needed, such very high Q resonators could be used for filtering and other rf signal control functions. A Q of about two orders of magnitude greater than conventional cavities seems to be achievable.

Another application of a superconducting cavity is in the generation of high peak pulse powers. A conventional microwave power source can be used to excite a resonator cavity for the storage of microwave energy. When the energy is released in a time interval much shorter than the storage time of the cavity, very high peak pulse powers can be generated. The work of the Soviet scientist Didenko* in 1980 described a microwave resonator system that produced a pulse power gain of 18 dB at a frequency of 3 GHz. A microwave generator producing 1.6 MW, 3 μs pulses delivered an output of 70 mw, with a pulse width of 15 ns, using a conventional (non–superconducting) resonator design. The efficiency of conversion of long pulse (3 μs) energy to short pulse (15ns) energy was 80%, but the overall efficiency for generation of the short pulse was 30%.

According to Didenko, much greater power amplification gains are possible if superconducting materials are used in the cavity design. This technique has promise for the generation of high power, narrow pulses, without the need for an accelerated electron beam.

14–6 PLASMA WAVE DEVICES

Chapter 7 discussed plasma concepts and also introduced the concept of longitudinal plasma waves. These waves occur in a plasma and have electric fields in the direction of their propagation. The ordinary EM wave used in radar and communication has perpendicular electric and

---

*Devyatkov, N.D., et al, "Creation of Powerful Pulses with the Storage of SHF Energy in a Resonator", Radiotekhnika i Elecktronika, Vol. 25, No. 6, 1980 (pp. 1227-1230).

magnetic field components that are normal to the direction of EM wave propagation. Figure 14-3 illustrates these conditions.

In Fig. 14-3a, the longitudinal wave in the plasma propagates in the direction $\hat{k}$. Its electric field $\bar{E}$ is along $\hat{k}$. In Fig. 14-3b, the EM "plane wave" has no components of electric field $\bar{E}$, and magnetic field $\bar{H}$, that are perpendicular to each other but lie in the plane x-y. The propagation is in the z-direction ($\hat{k}$-direction) which is perpendicular to the x-y plane. The longitudinal wave will not propagate in free space, at least in the EM wave sense. What it can do, however, is convert to an EM wave when the longitudinal wave strikes a discontinuity.

With the above concepts in mind, the problems in plasma wave amplifiers can be identified. The plasma wave exists only in the plasma and must be converted into an EM wave eventually. (The longitudinal wave of Fig. 14-3a must be converted into the EM wave of Fig. 14-3b at the device output port, and vice versa at the input port.) The amplification must occur in the plasma and thus growing plasma wave modes are also needed.

Plasma wave devices of the future can use one of two longitudinal wave modes that exist in the plasma. The modes have been called the optical and acoustic branches of the so-called dispersion relations. The dispersion relations in the plasma are found in theory by equating the longitudinal dielectric constant $K_\perp (k,\omega)$ to zero. There are two branches to the solution of $K_\perp (k,\omega) = 0$. These solutions give the natural (normal) modes of oscillation in the plasma. The lower frequency branch has wave velocities comparable to the acoustic wave velocity in the medium and is one in which the electrons and ions are able to move together cooperatively. The upper frequency (optical) branch permits oscillations that electrons perform that are too rapid for the ions to follow. In this optical mode the ions serve only to provide a neutralizing background. The optical mode is usually near to the electron plasma frequency.

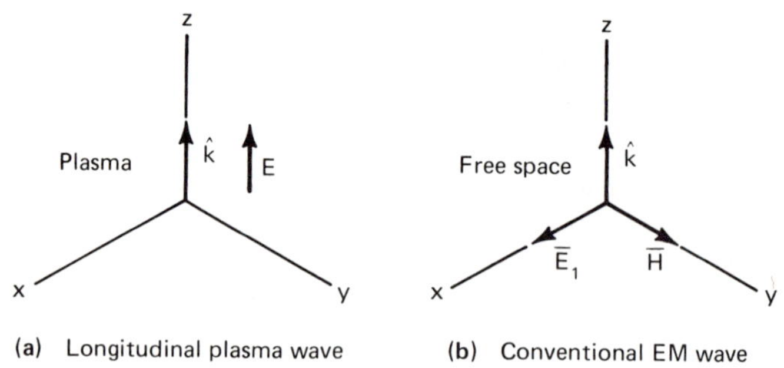

Figure 14-3: Comparison of Longitudinal Wave and EM Plasma Wave

Refering now to Fig. 7-10, the plasma amplifier must provide a means to couple efficiently an incoming EM wave into one of the longitudinal plasma normal modes. The mode must also be a growing mode and thus the plasma velocity distribution function must have a built-in growth mechanism such as those given in Fig. 7-5 and 7-6. Figure 7-5, with its lower velocity growing wave region, could support an acoustic mode. Figure 7-6, with its larger velocity optical growing wave region, could support that mode. Few plasma wave amplifiers have been devised and little success has been achieved.

The problems that must be solved can be tabulated. The following is a list of required capabilities:

1. Achievement of efficient plasma wave–EM wave mode conversion and coupling

2. Control of plasma wave mode growth rates

3. Maintenance of modest noise temperatures

4. Long term stability and reproducability of performance.

Plasma tend to be noisy and often can have electron temperatures higher than that of the ions. These non–equilibrium plasmas have some special characteristics, especially in the presence of growing plasma waves. Three temperatures can exist simultaneously in the plasma. (Electron, ion, and neutral gas temperatures can all be different.)

This writer is convinced that the first and second capabilities listed above are achievable, particularly the second capability. If the ion and electron temperatures can be kept near each other, the noise will be kept lower. Capability four presents a great challenge. The poisoning of emitter surfaces by ions and other contaminants will make reproducibility difficult.

Great promise lies in the development of very high power plasma wave amplifiers and oscillators. As evidence of this, foreign experiments in the achievement of very high power millimeter wave pulses using gyrotrons have demonstrated that the pulse power can be enhanced by two orders of magnitude by filling the gyrotron with a neutralizing plasma. (The plasma frequency was made less than the cyclotron frequency of the device to prevent mode shifting in the resonators.) The plasma removed the space charge (coulomb) forces that were limiting the electron beam drive current, thus permitting stronger gyrotron drive currents and higher power outputs. The self–healing nature of the plasma amplifying medium makes it attractive for high power, laser–like usage.

The device might also serve as a high power, incoherent noise source in the millimeter wave band. A triode of the type used in Fig. 13-2 when filled with plasma, might form very high EM pulse bursts. Flash x–ray devices have been formed in the past using high pulse plasma diodes. The phase and amplitude stability requirements (capabilities) would no longer apply to this plasma device application, making its achievement more reasonable. In this case the noise temperature could be made as high as practicable, as the desired output is reasonably pure Gaussian noise.

14–7 THE HYBRID DEVICES

14–7.1 The Electron Cyclotron Resonance Maser (ECRM)

The gyrotron tube tends to fill the gap between the conventional microwave tube and the laser. Figure 14-4 illustrates this tendency as it compares the peak and average power capabilities for these two types of devices. The laser is not satisfactory at long wavelengths as the energy in a quantum becomes too small. At wavelengths less than 10mm, microwave tubes become poor power generators as the energy density in a cavity becomes too large. The gyrotron does not depend on quantum steps and has a reduced power density due to its large (in terms of wavelength) cavity resonator. It, thus, is quite suitable for filling this unusual power gap in the EM power generation spectrum.

Since the gyrotron is a form of maser, it has equivalent population inversions brought about by the injection of electrons in a beam with high transverse energy. Nonetheless, it is a maser, which is close kin to a laser.

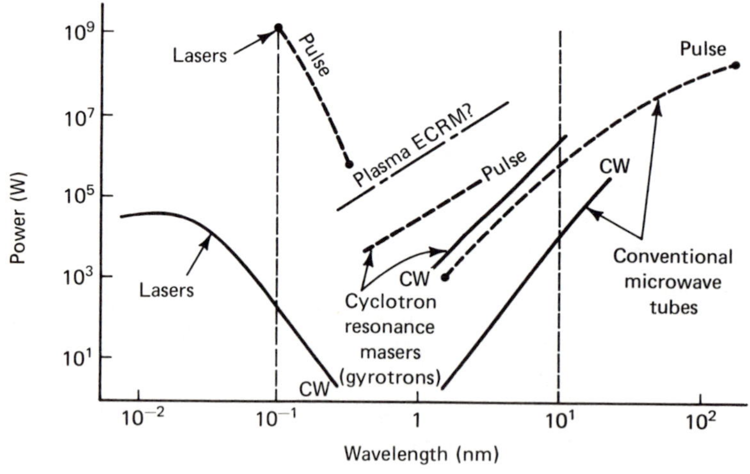

Figure 14-4: The Power Gap in Device Capability

What is suggested here is that a hybrid device concept might be devised that would use optical cavities and an electron cyclotron population inversion mechanism to help fill the power gap of Fig. 14-4. The gyrotron will be effective between 10 and 1.0 millimeters, but another device concept would be needed to extend from 1.0 to 0.1 millimeters. The new form of ECRM would permit optical–type resonator and amplifying medium designs. It would permit the multiple-path wave growth of a laser oscillator and have mirrors at either end of the cavity. It would also have spontaneous and stimulated emission in its amplifying (active) medium. It would also be desirable to not require superconducting magnets for its operation. The magnetic fields required for 1.0 to 0.1 mm cyclotron resonance are simply prohibitive, even with superconducting magnet technology.

Proper optical cavity design may also permit the direct coupling to a linear EM output mode, not requiring the higher order circular waveguide modes of the present gyrotrons. The present gyrotrons require complicated mode convertors to couple the antennas or other useful devices. Imaginative design based upon optical or quasi–optical techniques would permit high power generation in the device. It would also permit design of transmission lines that could handle this power at low loss.

14–7.2 The Plasma Cyclotron Maser

A form of ECRM may be possible that would be a hybrid between an ECRM and a plasma amplifier. When a magnetic field is applied to a plasma, new normal modes form in the plasma, some of which are close to the electron and ion cyclotron frequencies of these particles in the plasma, or are hybrid combinations of the plasma and cyclotron frequencies. This would be an extension of the plasma wave device concepts of section 14-6, but would involve different device geometries and wave growth concepts. An injected current system would now require Hall currents instead of the conventional field aligned Pederson currents in the amplifying medium. (This possibility is suggested in Fig. 14-4 by the Plasma ECRM line.)

In this type of hybrid device the plasma would not simply neutralize electron stream change, as in the plasma modified gyrotron. It would be an integral part of the bunching orbital mechanisms and enhance the efficiency of the hybrid amplifying mechanism. It would also permit higher power outputs if larger resonator designs were devised.

The presence of the positive ions in the plasma will provide an ion cyclotron orbital resonance in addition to the electron cyclotron orbital resonance. In addition, the electron plasma and ion plasma resonances are also present. Some of these may be coupled together in some type of pumping scheme via Manley-Rowe type mechanisms (see section 2–3 of Chapter 2). A thorough study would be needed to find the most promising new device. As a physicist would say, there are many degrees of freedom to this problem.

14–7.3 Solid-State Plasma Devices

There is an analogy between the plasma amplifier concept using a gaseous plasma and a device based upon a semiconductor plasma. The same theory of longitudinal plasma waves applies to either one, except that the loss mechanisms in the semiconductor are somewhat more severe. Coherent wave buildups tend to be broken up by scattering at the lattice sites. The concepts of Chapter 7 involving longitudinal wave buildups due to current systems in the plasma still apply. Velocity distributions that are non–maxwellian still provide proper growth mechanisms.

In a sense, the solid state equivalent of a plasma amplifier may have more promise. The stability of the plasma medium is assured in the solid state, and the temperatures may not be as prohibitive. They may also be produced in a smaller package, making them more attractive from that viewpoint. The ultimate of a solid state plasma electron cyclotron maser with applied magnetic field might also be considered here. However, it is confusing to try to combine all of these concepts at one writing. That will be left to the budding electronic designer as an exercise. The possibilities are vast and the imaginative designers are few.

It is this writer's hope that those who consider the possibilities here will seriously attempt to explore the new fields that lie ahead. Again, in the words of Solomon, "where there is no vision, the people perish———.''

Index

A

Abrupt junction (see step
      junction), 11
Absorption process, 131
Acceptor level, 10
AC Josephson effect, 154
Acoustic branch
      (plasma wave), 88
Alexanderson alternator, 2
Amplification process,
   of laser, 134
   of maser, 133
   of plasma, 87
Amplifier,
   four terminal, 67
   gyrotron, 169
   klystron, 116, 121
   laser, 146
   maser, 133
   noise power output of, 66,
      67
   parametric,
     double tuned, 23
     single tuned, 16
   plasma, 95
   reflection type, 69
   supercooled, 161, 163

Amplifier (*Contd.*)
   transistor,
     bipolar, 27
     unipolar, 31
   traveling wave tube
     (TWT), 110
   two terminal, 39
   TWYSTRON, 125
Angular frequency, electron, 168
Anode blocks, magnetron, 106
Applegate diagram,
   linear klystron, 117
   reflex klystron, 124
Aurora borealis, 82
Auroral simulation
      experiments, 90
Avalanche,
   breakdown, 50
   injection amplifier, 56
   injection device, 49
   shock front (TRAPATT
     mode), 58

B

Backward wave amplifier (BWA),
   principle, 108
   schematic, 110

Backward wave crossed–
    field amplifier
    (BWCFA), 110
Backward wave oscillator
    (BWO), 109
Backward wave principles, 108
Baking–in procedure
    (magnetron) 105
Band diagram (see energy
    diagram), 10
Barkhausen–KURZ (BKO)
    mechanism, 3
Barkhausen oscillator circuit, 5
Barrier injected transit time
    (BARITT) device, 60
Barrier injection device, 60
Bipolar junction transistor, 28
Bipolar microwave amplifier, 28
Bipolar vs. unipolar transistors,
    27, 37
Breakdown, avalanche, 50
Breakdown voltage,
    JFET, 35
    punch through, 52
Bolometer, with SQUID sensor,
    163
Brian Josephson, 151
Buncher cavity, 116

C

Capacitance,
    electrode, 2
    junction, 13
    tuning, magnetron, 107
Capacitor, variable, 13
Carbon dioxide (CO_2) laser, 136
Carcinotron (M—BWO), 111
Carrier,
    depletion, 11
    injection, 33
    velocity, 43, 54
Cathode,
    continuous emitting

Cathode (*Contd.*)
    Sole, 110
    injected beam, 110
Cavity "ringing" device, 194
Charge buildup,
Chemical laser, 138
Circuit stabilization, two–terminal
    device, 70
Circulator, microwave, 69
Coaxial cavity (Gunn
    oscillator), 47
Coherence,
    laser, 132
    maser, 132
Coherent pumping, 132
Collective oscillations, 87
Common base amplifier, 29
Common emitter amplifier, 29
Communications repeater
    chain, 80
Conduction band, 10
Conductor—definition of, 10
Continuous–emitting Sole
    cathode, 110
Continuous wave (CW) power
    output, 127, 166
Cooper pairs, 149
Coulomb forces—neutralization
    of, 96
Coupled–cavity devices
    (TWT), 119
Critical supercurrent, 154
Crossed–field amplifier
    (CFA), 109
Cryostats—definition of, 148
Crystaline solid state laser, 135
Current—voltage characteristic
    (JFET), 35
Cutoff frequency, transistor, 29
CW power of microwave and
    millimeter wave
    devices, 127
Cyclotron concept, electron, 168

Cyclotron frequency,
        magnetron, 99
Cyclotron resonance, 167
Cylindrical magnetron, 101

D

Debeye length, 81
Degenerate PARAMPS, 17
Delayed domain (LSA) mode, 44
Delay line, slow wave, 110
Depletion layer/region, 33
Device concepts, future, 188
Dielectric constant, 14
Dielectric permittivity, 83
Diffusion stabilization, 72
Diodes, solid–state, 9
Direct current (DC) Josephson
        effect, 153
Distribution function, velocity, 87
Domain growth, Gunn device, 44
Donor level, 10
Double drift IMPATT device, 54
Double resonant paramp, 23
Double tuned paramp, 17
Drain characteristics (FET,
        JFET), 34, 35
Drain current, 31
Dye laser, 137
Dynamic drain resistance, 35
Dynomic range, 74

E

Effective electron mass, 190
Efficiency, orbital, 171
Eigen modes (normal modes), 95
Electrodes, 95
Electrojet, plasma, 91
Electromagnetic spectrum, 130
Electromagnetic (EM) wave,
        interaction, 85
        propagation, 84

Electromagnetic wave (*Contd.*)
        reflection, 85
        refraction, 85
Electron,
        beam injection, 115, 167
        cyclotron resonance maser
                (ECRM), 166
        density, plasma, 83
        oscillations, Langmuir, 82
        paths, magnetron, 103
        thermal motion—tempera-
                ture due to, 68
Emission, spontaneous, 131
Emission, stimulated, 131
Energy band diagram, 10
Energy gap, 11
Equivalent circuit, Norton, 70
Equivalent circuit, Thevenin, 69
Extended interaction klystron,
        121, 127
Extended vacuum device, 189

F

Fast decay process, 134
Fast wave structures, 171
Fermi levels, semiconductor, 11
Field effect transistor (FET), 31
Five stage hybrid amplifier, 79
Flash x–ray devices, 182
Fluctuations, plasma, 86
Forbidden band gap, 10
Forward bias, 13
Forward wave crossed–field
                amplifier (FWCFA),
                110
Forward wave principle, 108
Four level laser, 134
Four terminal amplifier, 67
Free electron laser, 185
Frequency,
        electron cyclotron, 168
        plasma, 82

Frequency (*Contd.*)
    response, transistor, 27
Fusion experiments, 179

G

Gain—bandwidth product,
    solid–state, 79
Galium arsenide,
    diodes, 39
    FET, 27
    laser, 140
Gas laser, 136
Gate, semiconductor, 31
General amplifier, 67
Glass laser, 135
Gradiometer, SQUID type, 159
Ground state, 132
Group velocity, 109
Growing waves, 86
Gunn devices, 44
Gunn, John, 7
Gyrotron,
    amplifier, 169
    oscillator, 169
    principles, 166
    solid–state, 190
    vacuum, 171
Gyro–cyclotron resonance, 168
Gyro–klystron, 171
Gyro–resonance effect, 165
Gyro–TWT, 171
Gyro–TWYSTRON, 170

H

Hansen, W. W., 6
Harmonic content, 74
Harmonic generator,
    paramp, 22, 23
Harmonic gyro–frequency
    operation, 176
Heil, O., 6
Helix slow wave TWT, 118
Hertz, Heinrich, 1

Heterodyne mixing, 163
Highpower generators, 182
Hole and slot cavity, 104
Hole concept, semiconductor, 10
Holography, 147
Hot electrons
    (nonMaxwellian), 88
Hull, A. W., 5
Hybrid amplifiers, 79
Hybrid device, 125

I

Idler frequency, 17
Impact ionization transit–time
    (IMPATT),
    amplifiers, 56
    devices, 50
    oscillators, 52
Impurity, levels—linear
    junction, 15
Incoherent pumping, 146
Indium antinomide (InSuo)
    detectors, 194
Inductive tuning, magnetron, 107
Insights into the future, 188
Insulator—definition of, 10
Intense relativistic electron beam
    (IREB), 182
Intermodulation characteristic, 74
Intermodulation spectrum, 73
Intervalley transfer (GaAs), 40
Inverted population, 132
Ionosphere, 82
Ionospheric plasma, 82, 89

J

Josephson, Brian, 151
Josephson,
    detector, 156, 161
    oscillator, 154, 155
Josephson effect,
    AC, 154
    DC, 153

Junction breakdown, 50
Junction field effect transistor
       (JFET), 31
Junction transistor, 27

K

Klystron,
       bunching process in, 117
       extended interaction, 122
       multi–beam, 115
       multi–cavity, 122
       reflex, 122
       two cavity, 116
k–vector,
       concept, 87
       propagation direction, 196

L

Laser applications,
       future, 146
       present, 143
Laser— definition of, 129
Laser schemes,
       two–level, 134
       three–level, 134
       four–level,134
Laser types,
       chemical, 138
       crystalline solid–state, 135
       gas, 136
       liquid (dye), 137
       semiconductor junction, 140
Layer width, semiconductor, 14
Linear amplifier, 65
Linear beam (O–type) device, 115
Linear graded junction, 15
Longitudinal wave modes, 196
Low noise amplifiers, 65

M

Magnetron oscillator,
       baking–in procedure, 105

Magnetron oscillator (*Contd.*)
       cylindrical, 101
       planar, 101
       split anode, 101
       voltage tunable, 98
Magnetron resonance
       condition, 105
Manley–Rowe equations, 19
Manley–Rowe idealized
       circuit, 197
Marginal stability boundary, 92
M—carcinotron (M—BWO), 111
Metal–semiconductor field effect
       transistor (MESFET),
       32
Microwave amplification by
       stimulated emission
       radiation (MASER),
       133
Microwave circuit/beam
       interaction, 116
Microwave circulator, 69
Microwave—definition of, 130
Millimeter wave—definition of,
       130
Microwave transistor equivalent
       circuit, 29
Mobility of electron, 41
M–type device—definition of, 97
M–type
       backward wave amplifier
               (M—BWA), 109
       backward wave oscillator
               (M—BWO), 111
Multiple–beam klystron, 115
Multi–cavity klystron, 121

N

Nature, plasmas in, 81
n–channel JFET, 34
Negative differential mobility, 41
Negative differential resistance
       (NDR), 41

206   **INDEX**

Negative resistance
    magnetron, 101
Noise factor—definition of, 66
Noise figure, intrinsic
    (MESFET), 36
Noise power, 66
Noise temperature—definition
    of, 66
Nondegenerate paramp, 17
Noninverting paramp, 20
Norton equivalent amplifier, 70
n–type semiconductor, 10

O

Omega–Beta diagram, 108
Optical (plasma wave) branch, 88
Orbital efficiency, 171
O–type device—definition of, 114
Output amplifiers, 65

P

Parallel field (O–type) device, 114
Parametric amplifier (paramp), 16
Paramp inverter, 21
Phase velocity, 108
Pi–mode, magnetron, 104
Pinch–off (FET), 34
Planar magnetron, 99
Plasma cyclotron maser
    (proposed), 199
Plasma—definition of, 81
Plasma density, 82
Plasma frequency, 83
Plasma jet, 91
Plasma slab, 85
Plasma state in matter, 81
Plasma wave amplifier, 95
Plasma wave devices, 94
Plasma waves, 86
p–n junction (semiconductor
    junction), 11
P^+N N^+ diode, 51
Population inversion, 132

Potential well analogy, 118
Power conditioning, 187
Power gap, device, 198
Power generation efficiency,
    169, 173
Propagation windows, 180
p–type semiconductor, 11
Pumping,
    definition, for laser, 132
    definition, for paramp, 20
Punch through, 52

Q

Quanta—definition of, 130
Quantum electronics, 129

R

Radio frequency, 1, 129
Read–type diode, 52
Rectification, p–n diode, 12
Reflection amplifier, 69
Reflex klystron principle, 122
Reflex triode, 185
Refraction, electromagnetic (EM)
    wave, 86
Relativistic magnetron, 183
Resistance, low field, 47
Reverse bias, 12
Rotating space charge wheel,
    magnetron, 103
Ruby laser, 133

S

Saturation, laser, 134
Saturation velocity,
    semiconductor, 51
Self–healing, plasma effect,
    96, 197
Semiconductor—definition of, 10
Semiconductor laser, 140
Slow wave principle, 118
Slow wave structure, 118, 119

Sole plate, 111
Solid–state comparative
         ranking, 79
Solid–state gyrotron, 190
Split–anode magnetron, 101
Spontaneous emission, 131
SQUID, dc, 156
SQUID, rf, 158
Stimulated emission process, 131
Superconducting cavity, 149
Superconducting quantum
         interference device
         (SQUID), 156
Superconducting—definition of,
         149
Superconductivity principle, 149
Supercurrents, 151

T

Thevenin equivalent amplifier, 69
Third–order intercept point, 74
Third–order products, 74
Transconductance, solid–state
         device, 33
Transferred electron device
         (TED), 39
Transferred electron oscillator
         (TEO), 7
Trapped plasma avalanche
         triggered transit device
         (TRAPATT), 57

Travelling wave klystron
         (TWYSTRON), 125
Travelling wave magnetron, 102
Travelling wave tube (TWT), 118
Two–cavity klystron, 116
Two–stage noise factor, 67
Two–stream effect, 88
TWYSTRON schematic, 125

U

Ubitron, 193
Unipolar microwave amplifier, 31
Unit vector ($\hat{k}_1$), 94

V

Vacuum gyrotron, 167, 191
Vacuum triode, 3
Valence band, 10
Varian, Russel, 6
Varian, Sigurd, 6
Velocity distribution function, 87
Velocity modulation, 116
Velocity—time diagram, 117
Visible spectrum, 130

W

Wave angular frequency, 94
Wave functions (electron), 149
Waveguide (Gunn diode) cavity,
         48
Wave vector (k), 94